D0844817

DISCARDED
BALMORAL
CANADA

RODENTS

RODENTS

THEIR LIVES AND HABITS

PETER W. HANNEY

Illustrations by Jenny Agnew

DAVID & CHARLES
NEWTON ABBOT LONDON VANCOUVER

ISBN 0 7153 6667 X

Set in 11 on 13pt Garamond and printed in
Great Britain by Latimer Trend & Company Limited Plymouth
for David & Charles (Holdings) Limited
South Devon House Newton Abbot Devon

Published in Canada
by Douglas David & Charles Limited
132 Philip Avenue North Vancouver BC

CONTENTS

LIST OF ILLUSTRATIONS

Photographs without credits are by the author

IN TEXT

PREFACE

To many people, rodents are merely pests whose only claim to attention lies in the damage they cause; others, perhaps recalling stories of the beaver's sagacity or the lemming's suicidal marches, may take a more sympathetic view. Unfortunately for the reputation of the order as a whole, a few destructive species have loomed so large in human affairs that they overshadow the rest. No other order of mammals has been the subject of so many popular misconceptions, yet at the same time had the private lives of its members placed under such close scrutiny by biologists. Since there are so many kinds of rodents, with the most diverse and complex ways of life, it is not easy for even biologists to keep abreast of the many fields of study involving them.

In this book I have collated the observations of workers in many branches of biology, in an attempt to present a broad, general picture of the rodents and the reasons for their success, also to place into perspective some of the better-known facets of their lives. It has been impossible to mention more than a small proportion of the total number of rodent species. In spite of the voluminous literature about them, many aspects of their lives still remain a mystery. My main aim has been to show that rodents, whether in the wild, in a cage, or even stuffed in a museum, are far from mundane and are often more worthy of attention than more spectacular mammals.

Fig 1 *African water rat*, Dasymys

I

INTRODUCING THE RODENTS:
A Blueprint for Success

The rodents, without any doubt, are the most successful of modern mammals, with regard to numbers both of species and individuals. They flourish in most parts of the world, in almost every conceivable type of habitat. The exact number of species is unknown because classification is incomplete. Each time a group is critically examined by experts, it 'loses' a number of species. At present, about 1,650 species are recognised. Almost certainly, these will be reduced to nearer 1,600 within the decade, although a few species probably still await discovery. Whatever their actual numbers, rodents comprise about 40 per cent of mammalian species, and in every region they are the most abundant terrestrial mammals. The majority of species are rarely noticed by man, but they create a formidable force in influencing the environment, both in their everyday behaviour and as a source of food for carnivorous birds, reptiles and other mammals. A few have contributed in no small measure to human health and prosperity, while others rival man himself as agents of destruction.

An examination of the rodent's body gives few clues to the reasons for such phenomenal success. The order Rodentia derives its name from the Latin word *rodere*, meaning 'to gnaw', for the only unique feature is the pairs of large incisor teeth in the upper and lower jaws. Even in this, they are not greatly dissimilar from rabbits and hares. At one time, these were classified with the rodents, but, since they show several anatomical differences, including the possession of an additional pair of small upper incisors behind the first (Fig 2), they are now regarded as a distinct order, the Lagomorpha. The outer surface of the rodent incisor is coated with a hard layer of enamel, and each tooth is kept sharp and chisel-like by the grinding together of the upper and lower pairs. Unlike the teeth of all other mammals excepting the Lagomorpha, the rodent incisors grow throughout life. Being self-sharpening and everlasting, they can cut through all kinds of material, from fine grass to tough timber. Rodents have no canine teeth, and there is a space between the incisors and cheek-teeth called the diastema; this enables the animal to draw in the sides of its lips so it can gnaw without getting chips of material in its mouth. The only other mammal with a diastema and dental pattern which could be confused with that of a rodent, is the aye-aye, a rare lemur of Madagascar.

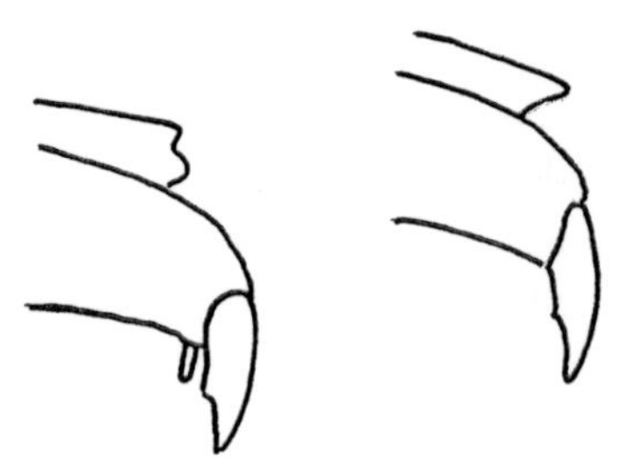

Fig 2 *Upper incisors of lagomorph (left) and rodent*

The rodents represent the second evolutionary experiment with teeth of this kind. An order of extinct mammals called the Multituberculata had rodent-like incisors and survived from the early Jurassic period, throughout the era of the dinosaurs, to the Lower Eocene, a time-span of some hundred million years. Rodents first appeared in the geological record at about the time when the multituberculates became extinct. If the rodents do as well as the other

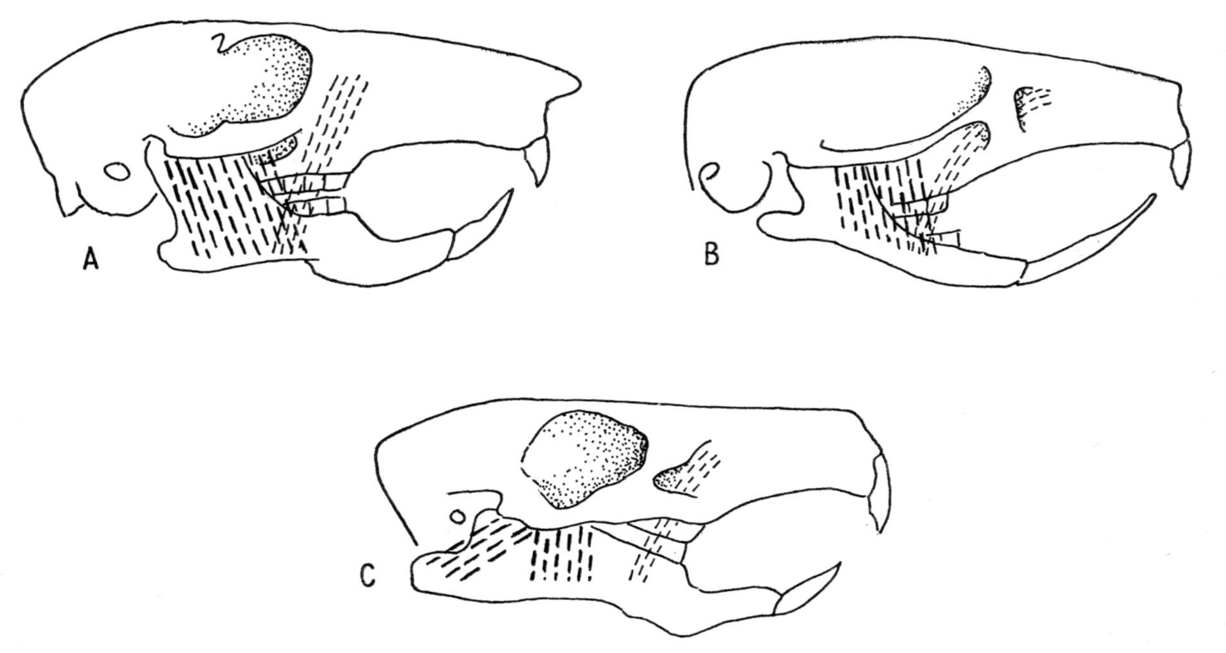

Fig 3 *Jaw musculature in the three sub-orders of rodents (the superficial muscle from the jaw to the lower side of the nose is not shown)*: (A) *sciuromorph (grey squirrel)*; (B) *myomorph (hamster)*; (C) *hystricomorph (guinea pig)*

incisor-bearing group, they should survive until the year AD 50 million.

The majority of rodents have three or four grinding or cheek-teeth in each jaw—a few have less, but only one species has more. The exception is the African mole rat, *Heliophobius argenteocinereus*, which has up to six cheek-teeth in each jaw. Frequently the sides of each cheek-tooth are deeply folded, and since the outer coating of enamel wears more slowly than the softer dentine, the grinding surface often forms a characteristic pattern which may be useful in classification. In some species, the cheek-teeth are rooted, but in others they are open at the base and grow throughout life. Apparently, all rodents lack a milk dentition. The jaws are loosely articulated to the skull, and when the cheek-teeth are used, the jaws are pulled back to disengage the incisors. This mechanism requires a complex muscular connection between the skull and jaws, and its evolution has been accompanied by three different skull modifications. These three types have been used for subdividing the thirty-three families of rodents into major groups or sub-orders. In the Sciuromorpha, or 'squirrel-like' species, the jaw muscle is attached to the cheek-bone as it is in other mammals; in the Myomorpha, or 'mouse-like' rodents, part of the muscle passes up from the jaw into the orbit and out through a hole to an anchorage at the side of the nose. The skulls of the Hystricomorpha, the 'porcupine-like' rodents, are usually characterised by massive cheek-bones, with an enormous hole just in front of the orbit—an arrangement which allows the muscle to pass straight up from the jaw to the sides of the face (Fig 3).

These terms refer to the skull rather than the animal's general appearance. Many 'squirrel-like' rodents look nothing like squirrels, for the sub-order includes pocket gophers and beavers. Similarly 'porcupine-like' rodents include most of the unique South American forms such as cavies, but also African mole rats and gundis. The 'mouse-like' rodents, however, generally live up to their name. They are the most successful, with about 1,100 species, compared with 365 sciuromorphs and 181 hystricomorphs.

Page 17 (*above*) American grey squirrel; (*below*) flying squirrel, *Glaucomys volans*

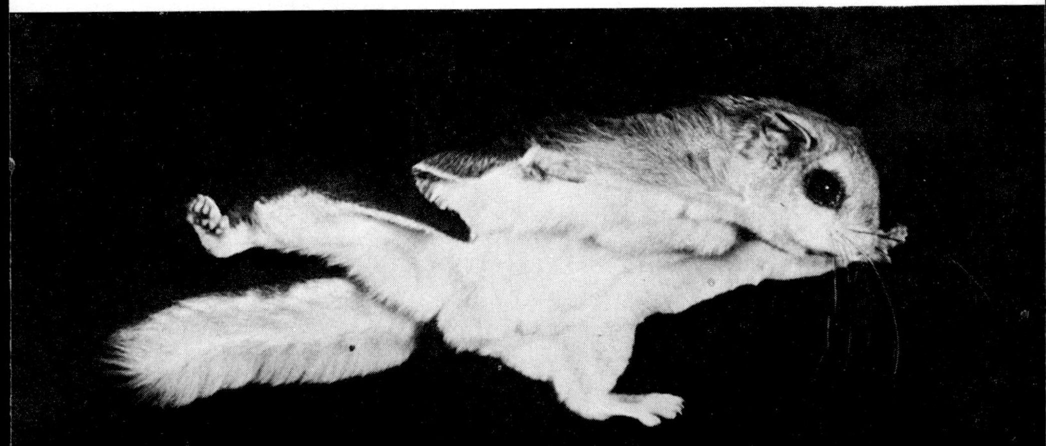

Page 18 (*left*) African climbing mouse, *Dendromus*; (*below*) African dormouse

The evolution of the rodents is a matter for speculation. The 'squirrel-like' forms are generally accepted as the most primitive, evolving first possibly in northern forests from where they migrated southwards throughout the Old World and through North America. They were unable, however, to establish themselves in South America to any extent. 'Mouse-like' rodents of cricetid (vole) type, probably evolved from 'squirrel-like' forms in the north and migrated south. In the tropics of the Old World, one line gave rise to the murid rodents. The Muridae, or Old World rats and mice, spread throughout the tropics to Asia and Australia, a few hardy forms managing to move north and establish themselves in colder climates. None of the Old World rats were able to reach the New World without the assistance of man.

It is difficult even to guess at the origin of the 'porcupine-like' rodents. Judging by the variety of forms alive today in South America, they must have arisen there, and the North American porcupine represents an offshoot which managed to migrate northwards. However, it is impossible to believe that some forms swam the Atlantic to establish the line in Africa and Asia, and it may be that the porcupines of the Old World are quite unrelated to those of the New, and the sub-order Hystricomorpha has been used merely as a convenient dustbin by taxonomists anxious to dispose of several Old World families whose affinities are impossible to determine. With such a complex order, there is often no means of knowing which characters represent true relationship, and which are the results of convergence.

Although feeding is a basic requirement of life, the blueprint for success is hardly the mere possession of two pairs of incisors, even if they are the most efficient in the animal kingdom. There are, however, no other remarkable features about the rodent's anatomy. The basic form consists of a long body, shortish legs, flat feet with usually five toes, and a tail. Shorn of its fur, the body structure is reminiscent of a lizard, or a member of the extinct synapsids which are considered close to the reptilian ancestors of the mammals. These, then, are the basic ingredients for the evolutionary spoon to stir and select: a highly efficient food-gathering mechanism and a simple, yet versatile body. Each stir can occur only when the sex cells divide and unite to form a new individual, and the rate of genetic change is largely dependent on the rate of reproduction and the life-span. Many rodents are extremely prolific. Mouse-like forms often produce several litters during an average life of six months. The procreative powers of some species are well illustrated by the golden hamster. This species was named in 1839 by a British naturalist who obtained a single specimen from Aleppo. It was not seen again until 1930, when a litter was captured by I. Aharoni of the Hebrew University in Jerusalem. All the hamsters in homes and research establishments throughout the world are descended from the male and two females of the original litter. In fast-breeding species, not only is genetic change rapid, but as Hershkovitz pointed out in his study of South American rats and mice, samples of specimens collected from the same locality at different times may vary so much that they appear to belong to different races rather than just different generations. Some of these differences are emphasised by the very early age at which some of them can reproduce. Variation may also occur in populations living very close to each other but occupying slightly different environments; those in habitats with plenty of food and few enemies are likely to be larger than those living in more rigorous conditions. Naturally, diversification is not limited to size, body measurements and other anatomical features. It also concerns life history and behaviour. It can be appreciated that in this situation, a museum specialist confronted by an array of variable specimens, has a difficult task in deciding whether they are members of the same species. In the heyday of natural-history collecting, between about 1850 and 1930, almost every population was given a specific or sub-specific name, sometimes even individual specimens were described and tagged with the name of the collector. Only recently have the thousands of described forms been reduced to a more realistic number of species. Chromosomes, blood and chemical studies are now used to supplement diagnoses made from measurements of bones.

It is clear that selection pressures by the environment on the simple blueprint (no doubt accelerated during the great changes brought about by successive ice ages) resulted in the evolution of the large number of rodent species we know today. Many orders of mammals are highly specialised and restricted to a particular mode of life. For example, seals and manatees are entirely aquatic; ungulates such as horses and deer are herbivorous running animals; insectivores and carnivores are generally dependent upon other animals as food; and flying lemurs are confined to forests. Rodents are not subject to such limitations. By a few modifications of the basic plan, such as lengthening the limbs, fusing a few vertebrae, and losing some fingers and toes, or a tail, some of them can outburrow the mole, outclimb the lemur, rival the seal in swimming and the hare in running. Like the birds, many species have mastery over two or more environments; some species may be equally at home underground, on the surface and in a tree, while others may spend part of their lives in water and part on land. Those remarkable teeth and jaws can cope with almost any food, besides being ready for other purposes.

Not all rodents are prolific breeders. Those with a slow reproductive rate have made little contribution to the number of species and are generally restricted to a particular type of habitat. The mountain beaver, for instance, uninfluenced by any selective pressures, has survived virtually unchanged for 50,000,000 years and is by way of being a living fossil. Species like this are now in danger of extinction through being unable to adapt to the environmental changes brought about by human activities. All evolutionary experiments have had their failures, and the rodents are no exception. It will never be known how many lines have become extinct, since rodent bones are notoriously fragile and easily overlooked. However, the examination of native middens in several Caribbean islands has revealed the bones of twelve genera of extinct hystricomorphs. One called *Amblyrhiza* had a skull of 400mm long and was probably the size of a black bear. Few living rodents have bodies more than a few inches in length. This factor enables them to live unobtrusively in

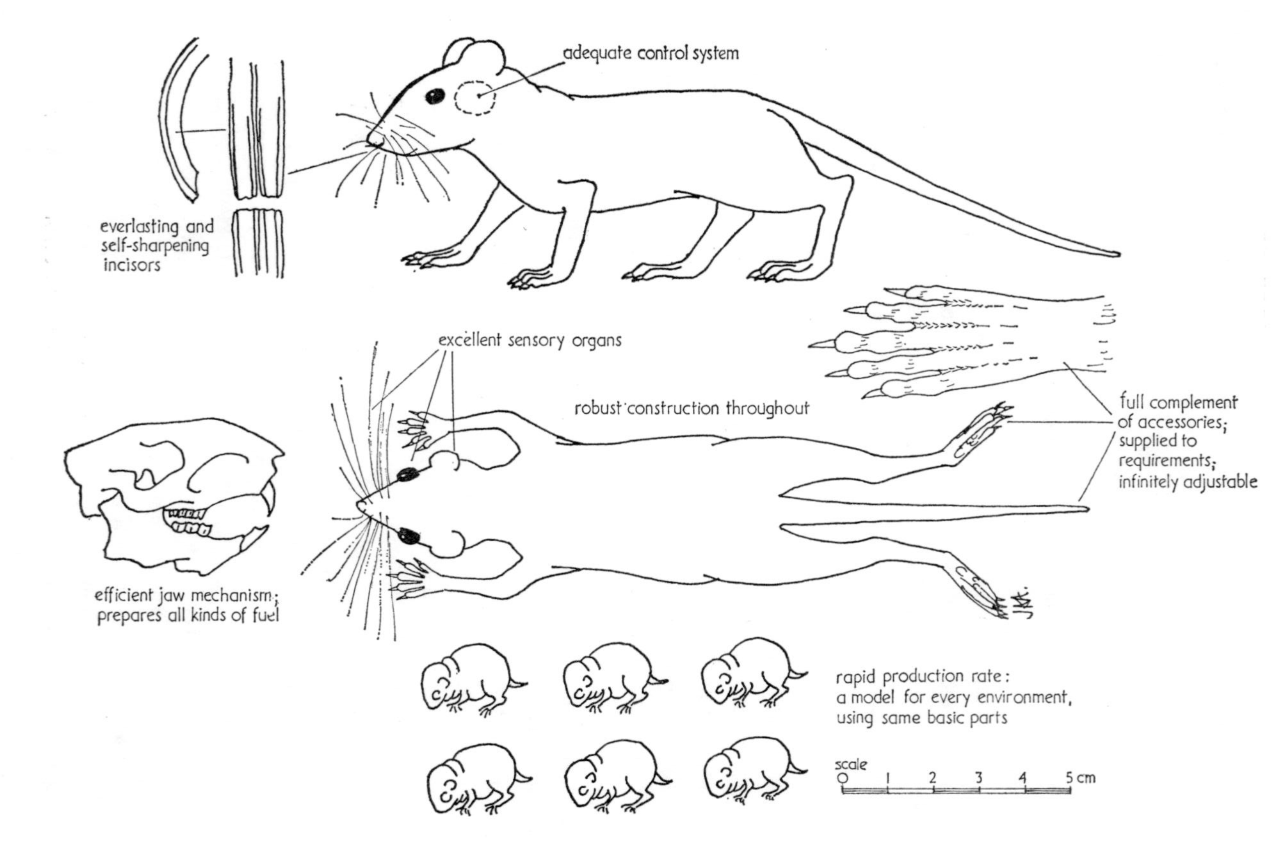

Fig 4 *A blueprint for success*

holes and odd corners, and also ensures they do not outstrip their food supply.

Little can be said about the rodent's brain, except that it has been subjected to the same evolutionary stirring as the rest of the body, and is reasonably large and unspecialised. The ability of a brain to correlate impressions of the environment and take appropriate action, depends largely on the number of sense receptors and effector structures. Most rodents employ all five senses and retain the use of all limbs and appendages. In addition, the chequered evolutionary history and multiplicity of habitats experienced by many species must involve the development of a great variety of mental processes.

Although rodents are found almost everywhere except in the oceans and polar regions, the distribution of species is uneven. It depends upon the number of habitats in a given area, environmental changes in the past, and whether the prevalent forms have a high rate of speciation. Collections of rodents made in different forests in Costa Rica and Panama never revealed more than 14 species,[55] 21 species have been reported in Uruguay,[5] and 43 in the whole of Malaya. On the other hand, western Europe has 52 species,[32] West Africa,[94, 131] while the Pacific States of California, Oregon and Washington can boast a total of 108 species belonging to 32 genera.[74]

2

THE CLIMBERS

The plant kingdom, in addition to providing animals with food and shelter, offers additional living space above the ground. An elevated habitat has several advantages over one on the surface, whether it is a few centimetres up a grass stalk or high in a tree. Fruit and leaves can be obtained before they fall, there is no danger from floods, and a high vantage point gives security from predators, although some species of carnivore can climb, only the martens have adopted an arboreal life. Branches and twigs also provide an effective shield against sudden attack by birds of prey. The early ancestors of man appreciated the benefits of life in the trees, and it is to them and the various selective processes involved, that we owe the blessings of our grasping hands and upright posture, as well as our 'slipped discs' and varicose veins. Man replaced the tree-top dwelling by the high-level apartment a long time ago, but a few groups of mammals, including some rodents, still live successfully in this ancient niche.

Life above ground can present problems. Rodents which live in trees tend to be more highly strung than those which live in burrows; they cannot afford to sleep too soundly for long periods in case the wind rises to a dangerous level. Pet flying squirrels become very nervous and restless during storms. The tree-dweller, if it is not to waste energy by continually trudging up and down, must be able to satisfy most of its needs in the branches, and the young have to be provided with a shelter of some kind. Tree holes provide homes for many species, others use old birds' nests or build their own. Nests are rarely made in the tops of trees but usually in the sturdier branches about half-way up. During summer, the American grey squirrel may use a nest built on slender branches, but in the windy, winter months it usually utilises one which is close to the tree trunk. In some species, the young are equipped with built-in security systems. Grey squirrels, two weeks old, have an instinct to grip and are able to support their own weight. The babies of some tree rats cling so tenaciously to their mother's teats that they hang like bunches of pink grapes as she runs along the branches. The arboreal rodent also faces the obvious hazard of an accidental fall, and various anatomical and physiological specialisations have been evolved to increase safety and climbing proficiency.

Members of about nine families of rodents are specialised to some extent as climbers; some rarely set foot on the ground in their entire lives, while others only climb to obtain food. The most primitive group, the sciuromorphs, or squirrel-like rodents, are primarily arboreal. Its members are distributed over most of the world excepting in the polar regions, Australia, New Zealand and the greater part of South America. Over the years, climates have changed and many forests have vanished. In such regions some squirrel-like rodents have adapted to a burrowing existence—they include the ground squirrels, prairie dogs and marmots. In most of the world's forests, tree squirrels are the most familiar wild mammals, not necessarily through their abundance, but because they are the most conspicuous. Unlike most small mammals, tree squirrels are active during the day and do not hide or flee from humans; in fact, they often display a marked inquisitiveness towards an intruder in their woods. In many respects, squirrels behave like birds, confident in their ability to move in three dimensions. There is little tendency for concealment and they are among the noisiest mammals, having a variety of twittering and chattering calls. No other mammals, excepting the monkeys, show such a range of colour and markings. Many are vividly marked in shades of grey, brown, red, black and white; in the African sun squirrel, *Heliosciurus gambianus*, no two individuals are alike. Unlike monkeys, squirrels apparently are unable to distinguish colours but their varied shadings play an important part in social life (page 149).

The squirrel's dexterity in the tree-tops is aided by the relatively long limbs, long flexible fingers and toes, sharp claws and the long, often flattened tail. The tail is important as a balancer when the animal is running along a branch, and is usually held over the back during feeding and resting. The Indian giant squirrel, *Ratufa indica*, which spends almost its entire life in trees, balances itself across a branch by hanging its head and forequarters over one side and extending the tail as a counterweight over the other. The tail also acts as a rudder and parachute, enabling the squirrel to make long leaps; *Ratufa* has been known to cross branches 7·6m apart. While leaping, most squirrels flatten the body by extending the legs and stiffening the tail; this slows the rate of fall and gives a tendency to glide. There is a record of a Mexican tree squirrel landing unhurt after leaping 180m down a precipice. Climbing specialisations are no handicap to running on the ground; a grey squirrel paced by a motorcar, reached a speed of nearly 29km/h.[146]

Anyone who has watched a squirrel's unhesitating progress through tree-tops will appreciate that all movements must be closely co-ordinated, and distances between branches must be estimated accurately and rapidly. The eye of the squirrel is large and has a rather flat cornea and lens to give a wide field of vision and a long focal length—features which, as every photographer knows, increases versatility. The retina, which may be compared with the film in a camera, is also specialised in its microscopic structure. The retina of the human eye usually contains two main types of cell known as rods and cones. The rods, which are highly light-sensitive and useful in poor light, are comparable to the chemical particles which allow certain films to be used indoors. In squirrels, and in many birds, the retina has no rods. It therefore has a low level of light sensitivity, but the uninterrupted series of cones give very fine discrimination, like a slow, grain-free film. In every vertebrate eye a blind area occurs at the point of entry of the optic nerve. In the diurnal squirrels, this does not take the form of a spot but a narrow strip, which runs horizontally above the centre of the eye. Since the image is reversed on the retina, the squirrel's upward field of vision is uninterrupted and no doubt this contributes to the animal's ability to run head-first down tree-trunks.

There are about 150 species of tree squirrels belonging to the family Sciuridae. They vary in size from the African pigmy squirrel *Myosciurus pumilio*, whose body is about as large as a human thumb, to the giant squirrels of India and SE Asia which reach 450mm in length. Most species feed principally on nuts and seeds, but often insects are eaten and many species are not averse to eggs or young birds. An Indian palm squirrel is particularly fond of the nectar of the silky oak, a tree frequently planted to shade tea plantations; while feeding, these squirrels become well powdered with pollen and serve as pollinators. The long-nosed squirrel (*Rhinosciurus*) of Malaya and Burma owes its unflattering sobriquet to the long, shrew-like rostrum which is associated with its diet of ants and termites. The upper incisors are small, but the lower pair are nearly half the length of the jaw and project well forward. Presumably they are used to lever away rotten wood to expose the insects; like many other insect-eaters, this species has a long and protrusible tongue.

Because of their attractive appearance and engaging ways, some species of squirrels have been introduced into countries far from their homelands. The American grey squirrel (page 17), a native of the broad-leaved woodlands in the eastern half of the United States, is the most widely ranging species in the world, due to the efforts of wildlife enthusiasts and landowners anxious to enrich their estates. Grey squirrels have been introduced to many parts of North America including British Columbia. In South Africa, Cecil Rhodes, whatever his shortcomings as an empire-builder, ensured a living memorial through the squirrels he liberated at Cape Town in about 1900. Those notorious animal lovers, the English, introduced grey squirrels to Britain on at least thirty-one occasions[146] between 1876 and 1929; by 1930 they had spread over nearly 25,000sq km and in

1970 were common in virtually every wooded district in England and Wales. There is some debate as to whether the charm of the British countryside has been enriched by the American newcomer; it has certainly been enlivened, for the woods have been echoing to gunfire ever since.

The red squirrel, *Sciurus vulgaris*, is the only species of tree squirrel indigenous to Europe, occurring in the forested zones of Europe and Asia from Britain to Japan. During the first two decades of the present century a drastic decline in the numbers of red squirrels in Britain was apparently caused by an epidemic disease. The disappearance of the reds from many areas of Britain left an empty niche for the newly introduced greys. At the present time, the red has been unable to regain from its larger rival the lost territories in the hardwoods, and it is generally restricted to conifer forests. In North America, three species of red squirrel which belong to a different genus (*Tamiasciurus*) are also mainly confined to softwoods. It seems very likely that the artificially created situation in Britain is ending as a repetition of the natural one in North America.

There is no doubt that squirrels affect the regeneration of woodland. One autumn morning, I estimated that 20,000 acorns were lying on my garden lawn; within six days they had all been taken by grey squirrels. The animals were far from hungry at the time, but in common with other squirrels living in cold and temperate zones, the drive to carry away and store food becomes intensified in autumn. In severe winters, squirrels depend almost entirely upon their stores which are sometimes buried, hidden in holes or in old nests. Frequently, especially during mild winters, the stores are neglected and the seeds germinate, often at a considerable distance from the parent trees. In this way the squirrel serves as an important agent in seed dispersal. The North American red squirrels feed chiefly upon the seeds of pine and spruce. The douglas squirrel cuts and fells the green unopened cones in autumn and caches them in groups of 160 or more, in damp places under logs or in small pits around the base of the trees. The dampness prevents the cones opening before the seeds are required for consumption. One observer in California watched a single squirrel cut 537 sequoia cones in 30min, removing each one by a single bite. In less than 3 days, all but 15 were hidden in the caches.[144]

So far as the tree squirrels are concerned, the adoption of an arboreal habitat has been an outstanding success. The specialisations involved have not interfered with ability to run, dig or otherwise adapt to changing situations, except that they are unable to see in the dark. Flying squirrels have taken specialisations a few steps further, but with rather less success. Thirty-four species belonging to eight genera are distributed in forests of the Old World, from Finland to China, and in southern Asia; two species occupy separate but overlapping ranges in North America. Instead of relying simply on outstretched limbs and tails to reduce the rate of descent, the flying squirrel has a gliding membrane on each side of the body, extending from the wrists to the ankles, and in some genera the membranes are continued to the neck and tail. The gliding membrane can be visualised as a fold of skin, consisting of a double layer of tissues covered by fur, and quite unlike the delicate wing membrane of the bat. Inside the fold, a rod of cartilage from each wrist serves as a spreader for the membrane. When expanded, the gliding membranes treble the area of undersurface, and not only reduce the rate of descent but allow the animal to glide at a very gradual downward angle (page 17). At least some species have aeronautical ability, at times riding on ascending air currents and being capable of banking. One species has been seen to use vigorous flapping movements to reach a point which was a metre higher than the launching place, and glide a horizontal distance of 135m. Unlike the tree squirrels, flying squirrels are nocturnal and little is known about their habits or distribution. Often their presence in an area only becomes known when a specimen is brought home by a domestic cat. The American mammalogist, Dr E. P. Walker, found that the southern species, *Glaucomys volans*, often shelters in tree cavities, where it hangs vertically by its fingernails or toenails. Before taking off, *Glaucomys* sways its head from side to side as if to estimate the distance to the landing-place. It launches itself with some force, and as the landing is approached, the tail is raised to give an upward lift and the limbs are extended to act as an airbrake. Immediately after landing, the squirrel scrambles to the other side of the tree to elude any possible predator.

There are no flying squirrels in Africa, but in the west of the continent their niche is filled by members of another family of rodents, the Anomaluridae or scaly-tailed squirrels. The scaly-tails superficially resemble flying squirrels and are also usually nocturnal in habits. They are easily distinguished by the two rows of overlapping scales on the underside of the tail near the base (Fig 5).

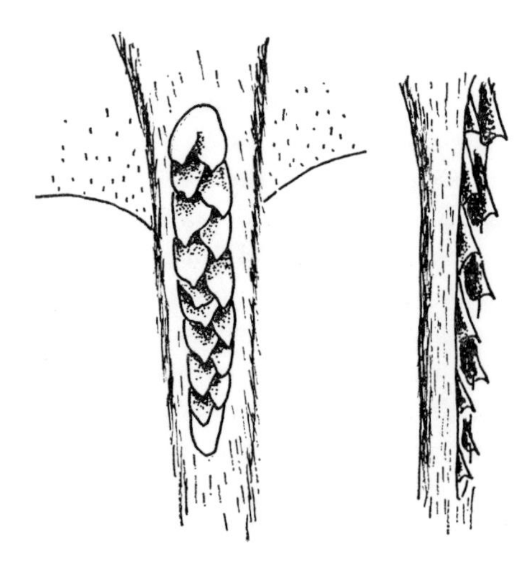

Fig 5 *Scaly-tailed squirrel,* Anomalurus*: base of tail*

These scales have spine-like keels which serve as anti-skid devices when the animal lands on a smooth tree trunk, also as climbing irons when it humps itself upwards in the manner of a looper caterpillar. The gliding membrane differs from that of the flying squirrel in that the spreader emerges from the elbow instead of the wrist. Some species are gregarious and share a den in a hollow tree; many museum specimens have been obtained by smoking them from such refuges. The pigmy scaly-tail congregates in numbers of a hundred or more. When their den is disturbed, they have been described as taking to the air in clouds, and 'floating away among neighbouring trees like bits of soot from a chimney'.[136] There is no evidence that scaly-tails make any kind of nest. Like bats, they simply use tree holes as a refuge, leaving the single young on a ledge inside and returning at intervals to feed it with well-masticated food.

It is interesting to speculate why, with their enhanced gliding abilities, flying squirrels and scaly-tails restrict their foraging activities to the hours of darkness, and have to employ such protective measures as cryptic colouration and silent movements. Probably, the gliding membrane interferes with the mobility of the limbs, and of course, any animal gliding at a modest speed of about 6·5km/h is inviting the attention of birds. By confining their activities to the night, they also avoid competition with tree squirrels which often share their territories. Daytime activity could present problems of water conservation and heat exchange owing to the extra body surface given by the membranes. Little work has been carried out on the vision of these animals, but the eye of the American flying squirrel is less specialised than that of a tree squirrel and possesses a blind spot. The reliance upon hollow trees and woodpecker holes for shelter must play a major part in confining these little-known rodents to mature forests.

The Gliridae, or dormice, is another family of climbers. Rather like miniature tree squirrels in appearance with their long furry tails and large eyes (page 18), dormice usually live in the lower levels of the tree canopy, in bushes and in rocky crevices. It is a small family with only seven species distributed over Europe, Asia and N Africa, and about five species in Africa, south of the Sahara. The English name is derived from 'la dormeuse', the sleeper—an appropriate description since all species become dormant during the most inclement season of the year. One South African naturalist realised just how apt the name was when he left a couple of rotten logs smouldering in the grate one night, and the next morning found a smoke-doped dormouse just emerging from a hole in one of them. Besides nesting in hollow trees, the African dormice use the thatch of houses, bunches of bananas, nests of spiders and various birds, including the mud nests of swallows. Chapin, in the notes of his Congo expedition, reported finding a female dormouse (*Graphiurus*) which had built her nest and raised three young in a compartment of a swallow's double nest attached to the underside of a rock. The only access was by walking upside-down for almost a metre on the nearly horizontal stone surface. At the side of this dormouse nest was a colony of paper wasps, and above it were several hundred earwigs.[66]

The edible dormouse, *Glis glis*, of Europe and Asia is the largest and most arboreal member of the family, living in the tops of 25m beech trees as well as in low bushes. Its claws are sharp enough to penetrate even the tough beech bark and allow it to run up and down the smooth trunks. An accomplished acrobat, it leaps between branches 7m apart, and sometimes hangs from a branch by its hind feet while trying to reach the beech mast. Falls are frequent, but the animals always spreads its legs and tail in squirrel-like fashion, and lands unhurt. In fruit-growing areas and in vineyards, the edible squirrel is sometimes a pest, having an insatiable appetite and invariably selecting the ripest and tastiest fruit. Ognev, with his penchant for accurate detail, recorded that in 10 weeks, 3 edible dormice consumed 272 cherries, 92 pears, 64 apples, 42 apricots, 58 plums, 25 grapes, 526 gooseberries and several hundred pumpkin seeds. With such a diet, it is no wonder that the species is so flavoursome to the human palate. Edible dormice usually build their nests in hollow trees but have been reported in the attics of houses. The winter quarters are in a rotten stump, a hole in the ground, or under a building. Like the common European dormouse which builds winter nests at the foot of a hedge, the edible species remains in hibernation through most of the winter. Some species appear to be semi-colonial, and if one nest is found it is most likely that others are in the close vicinity. In Israel, two or three nests of the forest dormouse sometimes occur in the same tree, and several more in adjacent trees.

Among the 1,100 species of rat-like rodents, there are many that show some climbing ability, most of them belonging to the Old World rats, the Muridae. Many rat-like rodents climb quite well without any high degree of specialisation apart from the long tail. The rat's tail, unlike those of squirrels and dormice, doesn't have to double as a parachute or blanket, and is unencumbered by dense fur. It is an invaluable balancing organ; black rats can enter a building by running along telephone wires less than 2mm thick. In performing such feats, the animal swings its tail from side to side in a figure-of-eight movement, and any injury or amputation of the tail interferes with climbing ability. Possibly this was in the mind of the farmer's wife whose experiments upon three blind mice have been familiar to small children for generations.

The long, slender tails of some climbing species can be curled loosely round branches and leaves to provide anchorage. One example is the pigmy climbing mouse (*Dendromus*) of Africa (page 18), which usually lives in low bushes and areas of tall grass. I became well acquainted with a member of this genus after being presented with a nest containing a new-born infant the size of a thumb-nail. After being weaned on milk and sugar solution, it went into a decline, reluctant to feed on fruit or proprietary baby foods. During one abortive feeding session, the mouse suddenly darted under a plate and emerged with a cockroach; after being efficiently decapitated, the victim was held like an ice-cream cornet, as its captor, which had never seen an insect before, munched away until only the wings were left. In the wild, the pigmy climbing mouse probably feeds mainly on grasshoppers, climbing plant stems to reach them. The tail is clothed with fine fur, all except the tip which resembles a flattened pink worm, and gently writhes and twists even when the rest of the tail is still. The tail tip is apparently a tactile area which saves the animal looking over its shoulder to see where its appendage can be twisted. A similar tactile area has been recorded for the tail of the prehensile-tailed rat (*Pogonomys*) of New Guinea, and probably occurs in other species. In nearly all tropical forests there are species of rats which are characterised by the possession of a stout tail which has the terminal portion devoid of fur or scales and possibly serving a similar sensory function. The tails of some climbing rats are covered with large, rasp-like scales which help to secure a grip when clinging to a smooth branch. In South America, where large tracts of land are covered by those giant grasses, the bamboos, several genera of spiny rats have become specialised for life in their evergreen canopies, grasping the shoots between the modified third and fourth digits.

In addition to the agile climbers which have to make rapid progress to avoid predators on the lower levels, there are the plodders, the heavy and ungainly rodents which lack the benefit of a whip-like tail, yet still manage to climb trees in pursuit of food. The North American porcupine is familiar to many people. Like heavy armour on nature's battlefield, it rolls on, secure in its protective coat; climbing trees by stretching its limbs round the trunk, reaching up with the right arm, then the left, then pulling up both hind legs simultaneously. The porcupine has been known to climb 18m in its search for leaves or bark. Various kinds of trees are attacked, particularly hemlock, red spruce and sugar maple, each animal having a strong predilection for previously visited trees. Some damage is caused, for the porcupine is a wasteful feeder, cutting off branches and leaving many leaves untouched. In some areas the waste clippings are a decided attraction for deer, and may help in their survival when there is little green vegetation at lower levels.

Unlike the Old World porcupines, all those of the New World can climb to some extent. The most adept are the tree porcupines (*Coendou*), which are found from Mexico to northern Argentina. Members of this genus have long flexible tails which curl upward like those of a few rat-like rodents. To aid in gripping, it has a callus pad on the upper side near the tip. The prehensile tail adds nothing to the tree porcupine's speed. It climbs so slowly and deliberately, it may sometimes be mistaken for a tangle of moss and leaves among the branches (Fig 6).

There is one even more unlikely climber than the porcupine. This is the mountain beaver (*Aplodontia*) which only occurs on the

Page 35 (*above*) Mountain beaver; (*below*) capybara

Page 36 (*above*) Coypu; (*below*) Australian water rat, *Hydromys*

western slopes of the Cascade range, from California to British Columbia. The mountain beaver (page 35) is difficult to fit into any scheme based on habits or habitats. It is not a beaver, nor is it confined to mountains, preferring to live in areas where the soil is saturated with water. Although it climbs, swims and burrows, it is not especially proficient at anything, yet somehow manages to survive. Its appearance resembles that of a muskrat with only a stub of tail. Although the pelt has been compared to 'a rug thrown out into the yard after a long winter's use, mouldering without dignity into the earth',[119] it was once used by Indians for robes and the animal probably acquired its other name of sewellel from the Indian name for these clothes. Its Indian name of chehalis is perpetuated as the name of a river and town in Washington State. The mountain beaver eats almost any kind of greenstuff and stores large quantities in underground larders, some of the food being cured outside first. During winter, the animal collects branches of evergreens, climbing

Fig 6 *Tree porcupine*, Coendou

to a height of about 7m by simply using the side branches like the rungs of a ladder. Its presence in an area may be indicated by a number of trees with bushy and uneven crowns caused by clipping. In one Washington plantation, 40 per cent of the planted trees were damaged by the animals.

The mountain beaver has two claims to recognition by the scientific world: it is the most primitive of living rodents; and its fur provides accommodation for the largest flea in the world, *Hystricopsylla schefferi*—a species which is over 9mm long. Perhaps it is by way of recompense that the mountain beaver carries no lice.

3

RODENTS OF RIVER AND SWAMP

No rodent is completely aquatic. The only fully aquatic mammals are the whales, dugong and manatees; all others must return to land to produce their young. Strictly speaking, rodents are either amphibious or semi-aquatic, but such terms make no distinction between those which spend most of their lives on land or in water. Most, if not all, rodents are capable of swimming in an emergency, and the general body form needs few modifications to produce a really efficient swimmer. There must also be many species which enter the water voluntarily in order to colonise new territories. Observations on a marked population of deermice living on islands in North American lakes showed considerable movement between different islands, but the mice studied swam only towards islands within their own vision. Lemmings, however, sometimes swim towards an empty horizon, like Vikings bent on reaching Valhalla. Some of these occasional swimmers are better equipped for immersion than others. The hamster, for instance, has been reported as inflating its cheek-pouches before entering the water, and using them like an improvised life-jacket. While the deermouse and lemming swim to reach new terrestrial habitats, other species actually live in water for part of their lives. There is great variation in the ways this habitat is used. Some species, such as the beaver, are completely dependent on water, while others, like the European water vole, can spend the whole of their lives on dry land.

A waterside home has much in its favour, and it is no accident that the earliest human settlements were on the banks of lakes and rivers. The fertile soil encourages plant growth, while the water itself may serve as a source of food, a more convenient means of communication than meandering trails, and a defence against any earthbound adversaries. These factors are as beneficial to rodents as to man, but the waterside habitat is not without hazards. The water which provides the means of life may also lead to its destruction. Floods can overwhelm homes and young in an instant, while drought may destroy all food resources and leave the water-dweller helpless and exposed to predators. Besides influencing the distribution of aquatic rodents, it is probably the tendency of inland water levels to fluctuate which has restricted and modified their evolution.

Although few zoologists would boast like the great Cuvier of being able to describe an animal from a single bone, it is usually possible to deduce an animal's habits and its habitat after examining its external features. Most aquatic rodents are recognisable by the thick fine fur, small ears and eyes, and also the broad hind feet which are often webbed. In the case of the capybara of South America (page 35), appearances can be misleading, for this giant among rodents reaches 1·3m in length, weighs up to 50kg, and looks like a cross between a dog and a guinea pig. The animal is actually semi-aquatic, and close inspection reveals that the feet are slightly webbed. The capybara lives among dense vegetation surrounding lakes and rivers, often wallowing in muddy pools and feeding upon aquatic vegetation. When threatened by hunters, it usually tries to elude pursuers by swimming underwater.

A much more familiar rodent, but with no special swimming adaptations, is the ubiquitous brown rat. Yet in many parts of the world it has adopted the waterside as its home. Sewers, as well as rivers, provide a rich source of food for this scavenger. Excrement, animal corpses and fishermen's bait are all acceptable.

Two genera of rat-like rodents have taken a few evolutionary steps towards aquatic specialisation with eminent success. They are the water voles (*Arvicola*), which are the largest European members of the family Cricetidae, and the African water rat (*Dasymys*). Both are similar in size and appearance, being rather thickset, with small furred ears and fine, dense pelage. They feed on a variety of aquatic and grassland plants but also take a small amount of animal food. The water vole ranges from the tundra of western Europe to central Siberia, and south to Israel and Iran. There are two major varieties: the smaller often lives in grassland and may never enter water in its life, while the larger form is usually closely associated with slow-flowing rivers and lakes. Even the larger, aquatic form can live independently of water. In times of flood and frost, it sometimes climbs small trees and bushes to gnaw the bark. Often, while walking the banks of the Avon at Stratford, I have alarmed water voles which were feeding in the hawthorns, and watched them plummet some 3m into the river below. The African water rat occurs through most of Africa south of the Sahara. It is normally seen in or near water, but on the high plateaux of Malawi, specimens have been trapped among rocks and bracken up to 1km from water; in lowland areas, all catches were made within 20m of a river or lake. Possibly there are two different forms, as there are of the water vole.

There are about eight genera of the Cricetidae and six of the Muridae whose members have attempted a far more aquatic life than the water vole or water rat. The majority have the hind toes webbed to some extent, or provided with a fringe of bristles which aid propulsion. Some have a swimming fringe, or keel, of short stiff hairs on the underside of the tail. Several genera are well adapted for an aquatic life. *Anotomys* of Equador, for instance, has no external ears to spoil its streamlined form—it is even able to close the ear orifices to keep out water. Seven species of South American fish-eating rats (*Ichthyomys*), in addition to the usual swimming modifications, show a most unusual development of the upper incisors. The outer corners of these teeth project downwards as sharp points and the rats are believed to use them like a gaff for catching fish (Fig 7A). Several genera of aquatic murids which inhabit parts of Australia and New Guinea have only two cheek-teeth in each jaw. These teeth are flat-crowned and usually large, presumably being used to crush the shells of water snails and other molluscs which form their diet. One New Guinea water rat (*Baiyankamys*) has two upper and three lower cheek-teeth on each side, and its incisors are identical to those of the South American fish-eaters (*Ichthyomys*). There is another species, *Mayermys*, living in the mountains of New Guinea, which is unique among both rodents and mammals in that it has but a single molar on each side of the upper and lower jaws (Fig 7). Since *Mayermys* has a slight membrane between the toes, it is presumably aquatic; but nothing is known concerning its feeding habits.

Most genera of small aquatic rats and voles live in the mountainous regions of the tropics—central and South America, New Guinea and Ethiopia—which remain relatively unexplored by mammalogists. These rodents are apparently rare even in their native haunts, several species being known only from a few skins in the world's museums. It seems that where small rodents are concerned, the adoption of an aquatic mode of life has not been particularly successful. Although about thirty small aquatic species have evolved, not one has a range approaching that of the water vole or African water rat. The restriction of so many species to upland rain forests is probably because it is in those areas alone that water levels are fairly constant, and there are few predators and competitors. As the rain forests are felled many of these highly specialised species will disappear through their inability to adapt to new environments.

Not all aquatic rodents are restricted to a few remote watersheds. If distribution is taken as the criterion of success, four of the most highly specialised swimmers are far from failures. These are the muskrat, Australian water rat, coypu and beaver. The muskrat, the largest member of the Cricetidae, has colonised most inland waters of North America, occurring from Alaska to Labrador, and south to Texas and Arizona. In Florida it is replaced by another species, the round-tailed muskrat.

The coypu (page 36) is a native of marshlands in South America. Like the muskrat, it is able to live in salt water as well as fresh. Both species have been introduced into many countries by fur-farmers, and as a result of escapes large numbers are now living wild in North America, Europe and Asia. Coypu have been deliberately introduced into the USSR, for the fur and as a source of meat. When the species was liberated in the southern republics during the 1930s, vigorous predator control was necessary to protect it against attacks from wolves, jackals, dogs and predatory birds. In its native haunts,

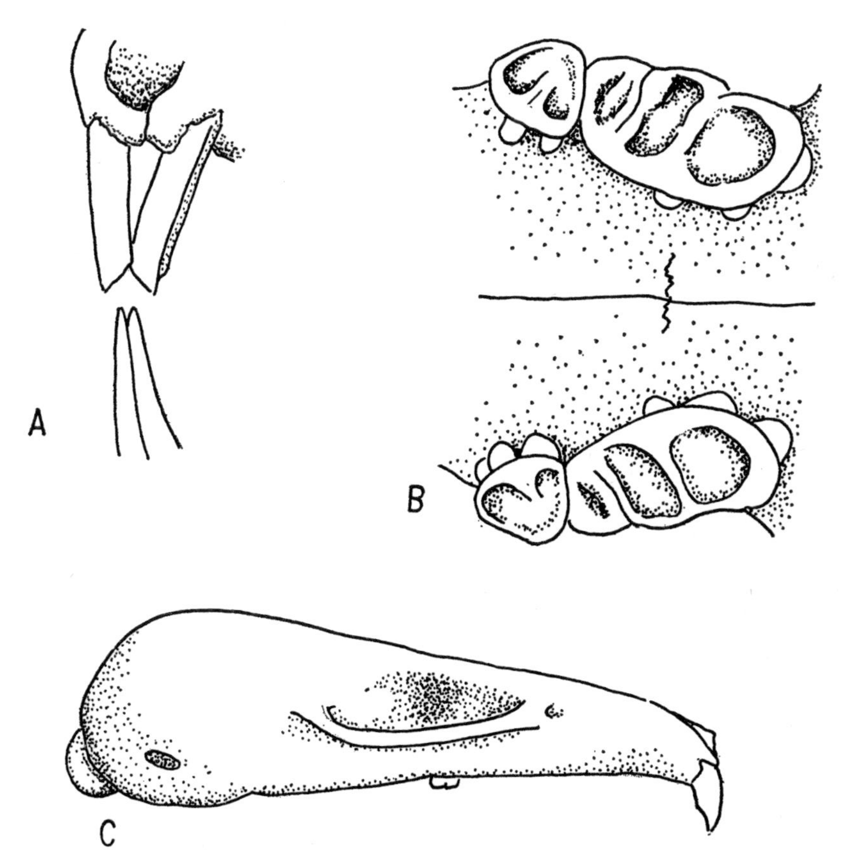

Fig 7 (A) *Incisors of fish-eating rat,* Ichthyomys; (B) *upper cheek-teeth of Australian water rat,* Hydromys; (C) *skull of* Mayermys—*a rodent with a total of four cheek-teeth*

however, the coypu apparently contends successfully against the caiman, jaguar and puma. Today, it seems to be more abundant in many countries of introduction than in some of its homelands, mainly through over-exploitation in the latter. Both coypu and muskrat feed chiefly upon aquatic vegetation, but sometimes eat shellfish. They are excellent swimmers, the coypu having four webbed toes on each hind foot, while the muskrat, which has only partial webbing, uses its laterally flattened tail as a rudder.

The Australian water rat (*Hydromys*) resembles the muskrat in size and appearance (page 36) but has a round, white-tipped tail. It is native to Australia, Tasmania and several islands in the East Indies. Like many of the smaller water rats in this region, *Hydromys* has only two cheek-teeth in each jaw and feeds mainly on invertebrates. It has an interesting method of dealing with shellfish. Immature mussels it simply crushes in its jaws, but older specimens, whose valves are too tough, the rat removes from the water and places upon a log or stone. Eventually the valves open in the sun, and when the rat revisits its feeding table, the succulent contents are ready for consumption.[162]

The beaver, until comparatively recently, was the most successful aquatic rodent. At one time, its range extended over most of the forested regions of the northern hemisphere; from Alaska to the Rio Grande, through Europe and Siberia, and south to the Mediterranean. Its decline in many regions over the last 500 years has been caused by persecution and the alteration of its environment through deforestation and other human activities.

These four rodents have many features in common, and are so specialised in their anatomy and physiology that they cannot exist away from water for any length of time. Young beavers for example, like capybaras, suffer from acute constipation unless there is water in which they can defaecate. Why are these species more successful than the small aquatic forms of the tropics? With the exception of *Hydromys*, they are predominantly vegetarian, and have no restrictions imposed by special dietary requirements; often they use water simply as a means of access to their food materials. Their large size is undoubtedly another important factor contributory to success. The beaver is second only to the capybara, while the coypu is about the sixth largest rodent. Not only does a large body deter many potential predators, but physiologically it is easily adaptable to a life in colder regions (see page 91).

By colonising the cold and temperate zones, sometimes with human help, these rodents avoid the hazard of drought, although they may face the threat of flood during the spring thaw. The young of aquatic rodents are especially vulnerable to floods, and in both the beaver and coypu the helpless stage is extremely brief. The newborn young are well furred and have their eyes open. Within a few hours they are able to move about, and the beaver can swim when only four days old. This rapid development, partly due to the long gestation period of around 120 days, is assisted by the very rich milk of the beaver which contains four times as much fat as cow's milk. The female coypu is able to nurse her young while in the water, for the mammae are well up on the sides of the body, just above the water-line when the animal is swimming. Although size contributed to success in the past, paradoxically it has led to the decline of all four species over the last few centuries. The soft pelts are all highly valued in the fur trade, the coypu providing nutria, and the muskrat musquash. One further aspect in the success of these species has been the evolution of complex patterns of behaviour, particularly important in the case of the beaver.

The beaver is the most popular rodent. Few people have seen the animal at all, least of all in the wild, yet its industry is a byword—and what schoolboy has not heard of its feats in felling trees and building intricate dams and lodges? The beaver earned the admiration of the American forest Indians for whom its fur and flesh provided the means of livelihood. The Indians always treated its remains with respect, the body being laid in a supposedly comfortable position with the hands, feet and tail tied to it before being committed to the water. If the body had to be eaten, the kneecaps were removed and ceremoniously burnt. Grey Owl, the half-breed trapper, whose writings in the 1930s did so much for beaver conservation, said that

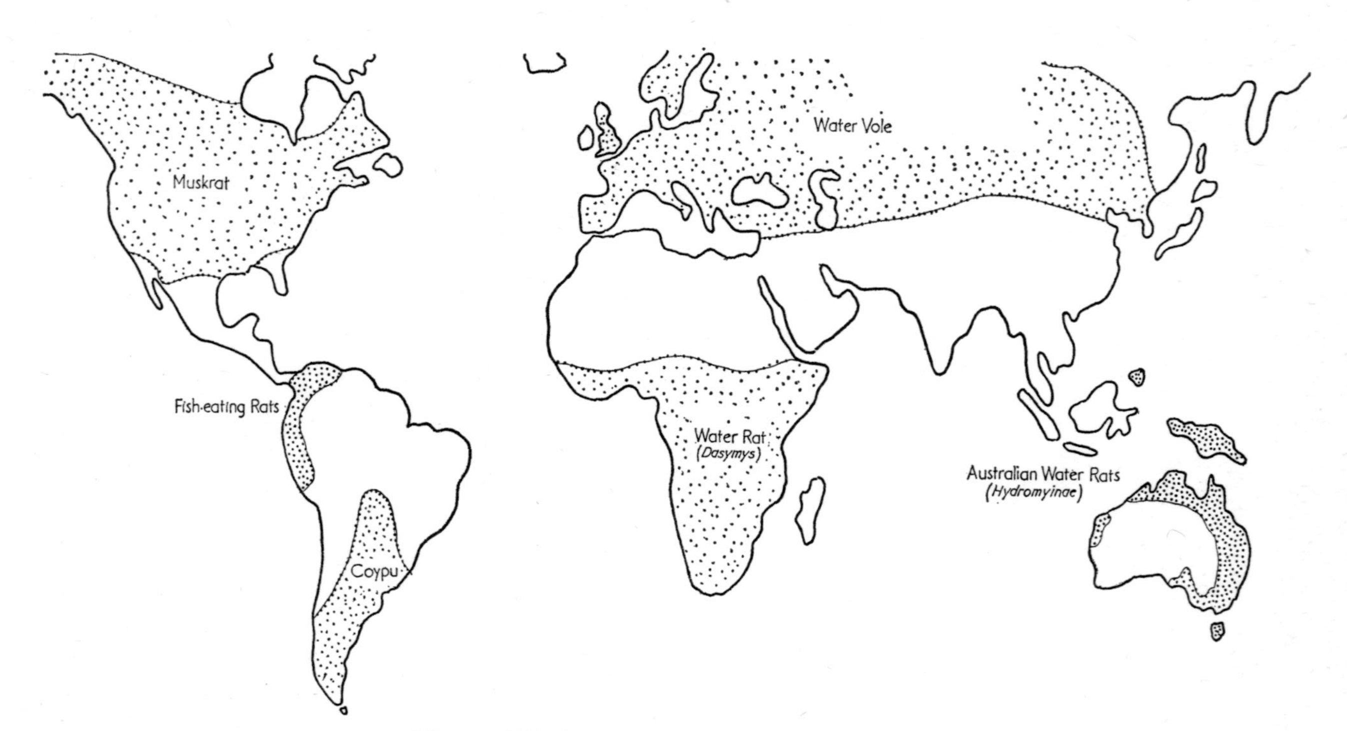

Fig 8 *The distribution of some aquatic rodents*

the Indian's high regard for the rodent was, 'Because they are so much like Indians'.

To a casual observer, the beaver seems to share many human characteristics. No other animal is able to control and influence its environment to such an extent. By means of its dams, water levels may be raised by as much as a metre and so completely transform large areas of countryside. In addition to its man-like constructional activities, the beaver lives in small communities, sits on its hind legs, uses its hands with great dexterity during feeding and building, and even mates like a human.

The sexes are impossible to distinguish visually: both male and female beaver have a similar type of cloaca and the male's testes lie in an abdominal cavity, a feature found in a few other mammals including elephants. While presenting no problems to the beaver, such sexual similarity does not help farmers and conservators attempting to rear the animals. Practised handlers palpate the abdomen to find whether a penis bone is present, but as the female has a similar bone, accurate sexing requires considerable experience.

The beaver shows many aquatic specialisations. It is quite streamlined in form, flattened in front and expanded behind. Like the coypu and muskrat, its fur includes a number of long guard hairs which matt together when wetted and trap a layer of air in the fine underfur. The eyes are protected by a nictitating membrane; the small ears are lined with dense fluffy hair which forms a waterproof seal, and both ears and nostrils can be closed by special muscles when the animal dives. The most conspicuous swimming adaptations are, of course, the webbed hind feet and the large flat tail. When swimming, the beaver holds its forelimbs still under its chin and each hind foot makes a rotating motion, more like a screw propeller than a paddle. As each expanded foot has a width of about 15cm, considerable thrust is exerted. During a straight swim, the tail undulates rather like a flatfish. In its horizontal position it acts as an elevator, but when turned vertically it becomes a rudder. By altering the position of its tail, the beaver can dive and turn very rapidly. Although so useful in the water, the heavy tail and broad feet only allow a waddling kind of progression on land. When alarmed, the beaver may lie pressed against the stream bed for up to fifteen minutes,[174] a time which approaches the average submergence period of many species of seal.

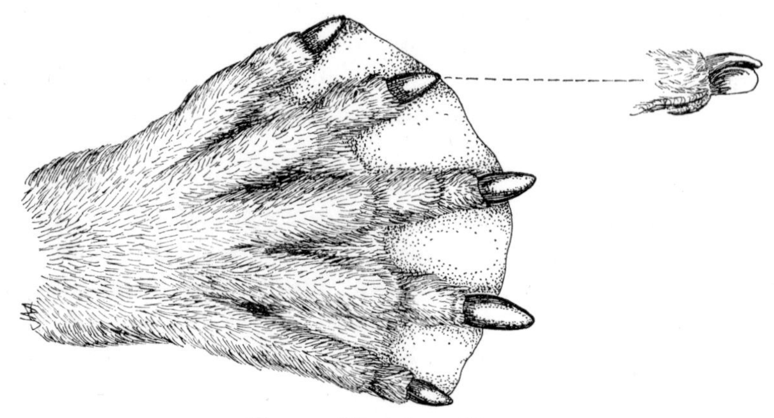

Fig 9 *Hind foot of beaver*

Anyone seeing a beaver, or a picture of one, sitting in a pleasant wooded glade during summer, might wonder how such a high degree of aquatic specialisation has evolved in an animal which, unlike the fish-eating rats, is not directly dependent upon water for its food. The beaver feeds upon almost any vegetable material; one study which was made in the USSR, showed that eighty-five different plant species were eaten.[114] During spring and summer it feeds on grass, nettles and any other green plants available, but during fall and winter its staple diet is fresh bark. Apart from conifer, almost any kind of bark is eaten, but birch and willow are preferred. The beaver's swimming and diving abilities are not, then, concerned directly with food gathering. Neither have they evolved simply to enable the animal to use the northern rivers as its chamber pot; on the contrary, the urge to defaecate in water may have arisen in order to guide the young beaver towards water, for in many parts of its range, its survival depends in large measure upon an aquatic environment.

It is in spring that the importance of the beaver's swimming ability can first be appreciated (page 53). At this time, the melting snows turn its habitat into a morass, where islands of vegetation are only accessible to an amphibious swimmer. During the floods, there is little danger from predators, but later in the year dwindling water levels could involve the beaver in long overland journeys when it would be highly vulnerable. The beaver is a large animal and faces the problem of over-cropping its food resources. Unlike large herbivores such as zebra and antelope, it cannot migrate or afford to extend its forays further and further from the water's edge. The beaver's well-known dam-building propensities overcome this difficulty; the dams contribute to its survival by raising water levels and so bringing water communications up to the food supplies.

It is largely the beaver's dam-building activities which have earned it a reputation for industry and sagacity. Beaver lore attributed the rodent with the skills and foresight of engineer, canal builder and lumberman. It certainly fells trees, digs canals and builds dams, but attempts to credit the animal with extraordinary powers of insight and reasoning can be dismissed, largely through the observations of Wilsson in Sweden, Wallace and Lathbury in Philadelphia, and various Soviet workers. The beaver's feats in controlling its environment are, however, no less remarkable for being the result of stereotyped behaviour patterns which have evolved by natural selection. There is no doubt about the animal's tree-felling ability, for the chisel-sharp incisors can sever a 12mm twig in a single bite, and are able to cut through tree trunks up to 50cm in diameter. The beaver has no control over the direction of fall. It usually stands on its hind-legs on the landward or highest side of the trunk, and gnaws as high up as it can reach; the tree normally falls down the bank into the stream, but in response to the forces of gravity rather than the skill of the cutter. When the tree is down, the smaller branches are removed and may be used in dam construction or stored as winter food. The beaver's canals, which are more like flooded slit-trenches, average half a metre in width and between a quarter and one metre in depth. They usually originate from simple trails made on boggy ground, the beaver converting them into canals by digging them deeper and flinging mud to the sides; the forefeet are used for digging while the webbed hind feet make excellent shovels. Near the lodge, the canals are often roofed and camouflaged with twigs.

Beaver dams are normally built across streams, never in large lakes. They average about 30m in length but may extend over 600m. Such large structures, which may be 3m high and 2m wide, are only possible because the beaver is a colonial animal where all members of the colony share the constructional duties. The sticks which form the dam usually lie parallel to the stream with the butts pointing upstream, just as they have drifted with the current. Wilsson, who kept beavers in a large aquarium, observed how they made the dam watertight by walking on the stream bed and deliberately kicking mud into the current—in due course the mud particles filled the interstices between the sticks. The stimulus for dam-building is given by the sound of running water. In nature, this is heard where water runs between rocks or over shallows; normally the building work continues until the sound can no longer be heard. With the scientist's usual lack of sportsmanship, Wilsson once placed a water pipe under a dam. When the beavers heard the water gurgling through the pipe, they built up the part of the dam just over it, until it eventually resembled a kind of hump-bridge. The dam-building drive was released by tape recordings of running water, also by the sound of an electric razor.

Wilsson's beavers demonstrated conclusively that dam-building is inborn, when an excellent structure was made by young animals which had never seen a dam. One of Wilsson's animals was even induced to build a dam in a tub of standing water when the water recording was played—a sound the animal had never heard before. In the wild, young beavers copy their parents to some extent, especially in the finer points of dam-building. Just as human populations have distinctive styles of craftsmanship, so each beaver colony or group of colonies may make a recognisable style of dam.

The beaver colony is seldom content with a single dam, but usually builds a series which progresses outwards from the lodge and extends the limit of navigation. One would expect to find maximum work during summer when water levels are low, but in Philadelphia Wallace and Lathbury found that little work was done at this time. Instead, there was feverish activity in spring when levels were at maximum. These American researchers found their beavers to be far from rational in their building programme, often repairing badly sited dams and allowing more important structures to disintegrate. Although so much time is wasted upon useless dams and by duplication of effort, dam-building must be sufficiently effective in improving water levels because otherwise natural selection would discourage the trait. Wallace and Lathbury suggest that the main purpose of dam-building is to provide maximum water round the lodge during the reproductive period. High water levels at this time must contribute to the survival of the young. However, since other workers report an increase in building activity during the fall, at least in Russia, it seems that the building drive may be influenced by factors which vary in different regions.

It is in winter that the survival value of the beaver's aquatic adaptations are really apparent. At this season its habitat is transformed; covered by snow and ice, it is readily accessible to wolves, lynxes and other predators, and with its dark fur and strong odour, the beaver would be at risk whenever it left the lodge or ice-covered water. Besides protecting the animal, the water also preserves the food stores which were collected during the fall. Ognev reported that along some beaver trails in the USSR willow branches are so neatly stacked at this time that they resemble man-made woodpiles. Later the branches are put in the stream and carried near the lodge. They are then stuck in the mud or woven together to prevent drifting, and eventually a large food pile may be built up. Ognev estimated that a single beaver needs nearly 4cu m of timber for its winter food supply, and Wilsson recorded one food pile which was 16m long and contained 90cu m of material.

A frozen river must be one of the safest places imaginable for a food store, but there are difficulties to overcome if the consumer is not to drown, or at least imbibe vaste quantities of water with every meal. Fortunately, the beaver is able to gnaw and feed when submerged. The upper lip is deeply cleft and both sides can be folded behind the incisors to make a watertight seal. The animal can breathe normally when on the surface, even if the mouth is opened underwater. Water does not enter the lungs because the front end of the windpipe, the epiglottis, lies above the soft palate. The epiglottis has to be retracted to allow swallowing, but only after the lips are closed. During underwater feeding, water need not be swallowed because in addition to the lip-seal at the front of the mouth, an elevated part of the tongue can be pressed up against the palate to make another seal at the back.[31] Whales also have an epiglottis lying above the palate, but here the position is permanent.

If the dams ensure a good depth of water during severe winter weather, the beaver can travel underwater between its lodge and foodstore, so avoiding the risk of terrestrial predators as well as frostbite. Owing to naturally fluctuating water levels below the ice, there is usually an air space which is sufficient for the beaver to breathe. Wilsson watched his animals deliberately make a furrow in their dam to liberate water and create an air space.

There is one other curious feature of the beaver's anatomy which may be attributed to its mode of life. As the animal spends most of its time in water or in a damp lodge, one would expect its skin to be a veritable hotbed for micro-organisms. In fact, it does suffer from severe infestations of small ticks, *Schizocarpus minguandi*, and is probably unique in accommodating a species of beetle in its fur; no one is quite sure whether the beetle feeds on the ticks or on the beaver's skin. In order to alleviate the tick problem, the beaver is provided with a couple of tick extractors. The claw on the second hind toe is divided; the upper half is normal but the lower consists of a horny plate with a skin-like covering. The two halves are movable, and ticks are removed by dragging the hair through the pincer (Fig 9). Beavers are often seen grooming each other, and it seems that in tick extraction, as in other aspects of life, communal living has its advantages.

Page 53 (*above*) Beaver; (*below*) field vole, *Microtus agrestis*

Page 54 (*above*) African harsh-furred rat, *Lophuromys*; (*below*) Mara

4

LIFE ON THE SURFACE

As the earth's surface is crowded with a host of mammals, birds, reptiles and invertebrates, some rodents have opted for less competitive environments in trees, fresh water or under the soil. The overwhelming majority, however, manages quite well on the ground. Such a variety of habitats are available that terrestrial rodents have diversified to exploit each niche to the full. It is not surprising that about 80 per cent of all rodent species are primarily terrestrial, although most of these can also burrow and many are able to climb or swim. Usually the specialisations towards each surface niche are less discernible than, say, the webbed toes of a swimmer or prehensile tails of climbers, but the number of adaptive permutations is incalculable. Sometimes they involve internal anatomy and behaviour rather than external appearance, yet among the 1,400 or so surface-dwellers it is possible to recognise several common adaptive trends.

The distribution of surface-dwelling forms increased during the Tertiary Period, when aridity in many parts of the world led to the replacement of forests by grassland, steppe and desert. In the grasslands, many rodents became more specialised in their feeding habits; instead of devouring anything edible which came their way, they evolved the most efficient means of cropping and digesting grass alone. After all this was the most plentiful food material and no energy had to be wasted in searching for it. Unfortunately grass is not an ideal type of food—it is hard on the teeth, difficult to digest, and large quantities are needed for a satisfactory meal.

The most specialised grass-feeders have high-crowned cheek-teeth which have a number of complex folds. Those of some forms are so close together that each set has the appearance of a single elephant molar, and the large rasp-like surfaces are able to shred the harshest grasses. Such is the wear on the teeth, that many of those superb mowing-machines, the voles, have cheek-teeth which are rootless and grow throughout life. It might seem impossible for a vole to be able to wear out a set of rooted teeth during its life-span of often less than a year, but it is doubtful whether the molars of any other mammal are so overworked. The social vole of Asia, for example, needs three times its own weight of grass every two days. Ognev recorded that in a year, ten captives had to eat either 73·5kg

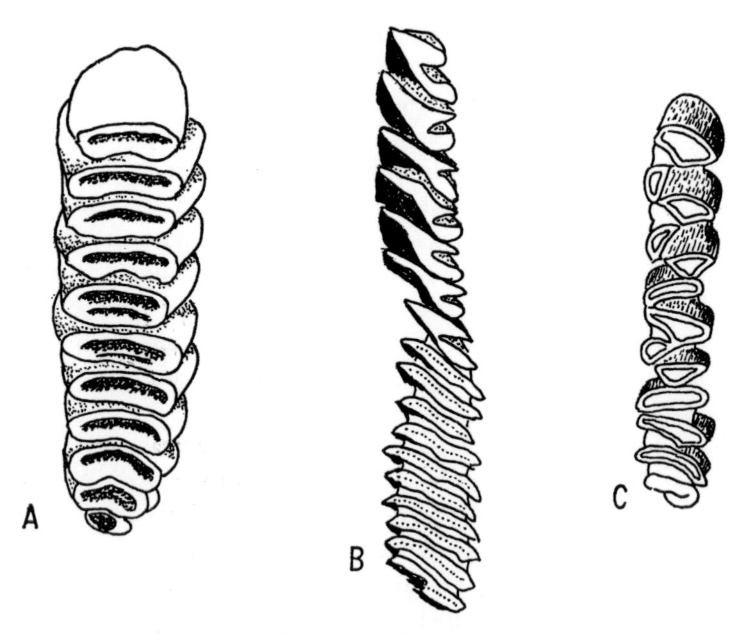

Fig 10 *Cheek-teeth (left, upper) of grazing and browsing rodents:* (A) *African swamp rat;* (B) *capybara;* (C) *wood lemming*

of greenstuff or less than half that weight of barley—no mean feat for grinding surfaces whose total area is smaller than that of a child's finger-nail. Most herbivorous rodents are active at any time of day or night. Owing to the relatively large amounts of fuel needed by the bodies of small mammals, those such as voles and lemmings whose appetites aspire to nothing more nourishing than grass, are doomed to spend a substantial part of their lives just eating. Such forms are generally confined to areas where vegetation is available through most of the year. Some species lay up food stores, but none of them can build up sufficient fat reserves in their bodies to allow them to hibernate. Larger grazers, such as marmots and prairie dogs, do not suffer from this limitation.

The digestion of large quantities of vegetation needs a sizeable stomach and a long intestine. Cellulose is also a difficult substance to break down and all herbivores possess a specially modified part of the alimentary system which serves as a fermentation tank where digestion can be assisted by micro-organisms. In cows and camels, freshly cropped food is swallowed and passed through part of the four-chambered stomach where it is partially broken down by bacteria. It is then regurgitated and subjected to a thorough chewing before being swallowed a second time and passed to other chambers of the stomach where digestion is completed. Rodents lack such an elaborate type of stomach, but fermentation can be carried out in the caecum, a pocket of the gut which is equivalent to the human appendix. Some kinds of rodents seem to be unable to regurgitate and their inability to vomit makes them particularly susceptible to certain poisons. To secure a return journey for the food, they are forced to adopt the simple expedient of eating their faeces. This habit, known as coprophagy, has been recorded in many species of rodents as well as in the rabbits and hares. The hare's indulgence in the practice earned the disapproval of Moses in the often-quoted Book of Leviticus. Coprophagy is essential for the well-being of herbivorous species; for example, house mice fed on greenstuff die within a few weeks if coprophagy is prevented. After the food's first passage via the caecum, the faeces are soft and rich in vitamin B_1, and their consumption not only ensures adequate digestion of food, but also the retention of valuable vitamins and water, subsequent faeces being hard and fairly dry. To some species, the first faeces are apparently more appetising than any vitamin pill; they are so relished by young beavers that the animals roll back on their hind legs and lick up the green substance directly it leaves the anus. Each newborn herbivore needs a culture of bacteria in its caecum before it can be weaned; in most cases this is probably obtained from the faeces of the mother. Guinea pigs lick this food from the mother's anus soon after they are born. Besides ensuring that maximum benefit is derived from the food, coprophagy allows the animal to fill its stomach as quickly as possible, then digest the food in the security of its shelter or burrow.

As a rule, grass-eating rodents make complex networks of runways through vegetation, small heaps of droppings and grass cuttings often betraying their identity. Many species bite off a convenient length of grass, hold one end in their hands and pass the other through the rapidly nibbling incisors, discarding the butt end and picking up a fresh length, rather in the manner of a confirmed chain-smoker. In its narrow runway, the grass-feeder is almost as insulated from sound as a burrowing species. As acute hearing is unnecessary, the ears are small and do not impede the progress of the low body between the plant stems. The tail is also reduced in length, but even voles can climb and jump when hard-pressed; European meadow voles (page 53) can easily leap a 20cm wall and climb a 50cm wire fence. In the northern hemisphere, voles and lemmings are the principal grazers, but in the south their niche is filled by several different types. Guinea pigs are the major grass-eaters in some parts of South America. In Uruguay they are the commonest rodents, sometimes being so abundant that the grasslands are seriously depleted. Besides the twenty species of guinea pig, there are other South American grazers, including the coney rat (*Reithrodon*) which has grooved incisors and a large caecum. In Africa, this genus is paralleled by the swamp rats (*Otomys*) which are like large voles in appearance—the molar teeth, although rooted, have a large number of laminae. The largest grass-eating rodent in Africa is a species of cane rat (*Thryonomys*) which frequently attains a body length of 60cm.

Seed-eaters and general feeders face fewer digestive problems than grazing forms. Although they lack a constant supply of any single kind of food, they are able to change their diet according to what is available. While it is usual to find only one species of grass-feeding rodent occurring in an area, several species of omnivore often share the same habitat without competing. Generally they are not restricted to runways, and as they have to search for food they are extremely agile and many have long tails and well-developed ears.

Deermice (*Peromyscus*) are among the most successful small rodents of North America. About fifty-five species occupy almost every conceivable habitat from forest to desert. Each species is far from restricted in distribution, for an incredible number of races or subspecies have evolved to fill different niches. The long-tailed deermouse, *P. maniculatus*, for instance, which is widely known as a laboratory animal, had sixty-six sub-species at the last count. Many studies have been made on the feeding habits of deermice, and usually the diet shows a marked change through the year: *P. maniculatus* in Indiana, for example, feeds on wheat seeds and soybean in winter and on moth and butterfly caterpillars in the summer.[171] Sometimes closely related species are found to be sharing the same habitat, but close study normally reveals some difference in food preferences. In Colorado, the golden-mantled ground squirrel, *Citellus lateralis*, and the least chipmunk, *Eutamias minimus*, live together, and in summer both feed on the common dandelion. But while the chipmunk prefers flowers and seed heads, these are rarely touched by the squirrel, which generally eats the stems.[126]

Because of the damage some rodents cause to grain, crops and plantations, it is sometimes believed that the Rodentia is an order of vegetarians. In fact the majority are omnivorous, feeding on almost anything edible, including animal tissues. Several species are entirely carnivorous. Weight for weight, animal tissues have far greater food value than plant material, and in some regions, insects and other invertebrates may be the only food available for part of the year. Rice rats (*Oryzomys*) living in the salt marshes of Georgia, prey on insects and small crabs in summer, and captives fed on animal food gained weight more quickly than those on a vegetable diet.[142] In Uruguay, Barlow examined the stomach contents of twenty-two species of rodents and found that only six fed entirely on vegetation. The twenty-six African species which I examined in Malawi included only four total vegetarians, all the others ate some insects, while five fed almost entirely on small invertebrates. The harsh-furred rat (*Lophuromys*) was found to feed mainly on worms, and its sharp claws are useful in subduing its slimy prey. Captives died within a week unless they were fed with worms, insects or frogs. The South American swamp rat (*Scapteromys*) feeds on worms too, and also has long claws. Both the South American burrowing mouse (*Oxymycterus*) and the North American grasshopper mouse (*Onychomys*) are almost entirely carnivorous, and their stomachs, like that of the harsh-furred rat, have specially toughened linings to cope with the harder parts of the prey. Just as leaf-eaters can be diagnosed by the laminated and flat-surfaced molars, confirmed carnivorous rodents generally have well-developed cusps on the cheek-teeth; in some, catching efficiency may be increased by the possession of forward-projecting incisors. The Congo link rat, *Deomys ferrugineus*, has cheek-teeth which resemble those of insectivores, having long, sharp cusps (Fig 11). Little is known about the habits of this species, but it has been observed wading in forest streams to catch crustaceans. It is not often that an animal which fills the stomach of another is able to leave a record of its demise, but in 1958 Oliver Pearson donned the cape of Sherlock Holmes when he discovered that a diet of cochineal insects imparted a permanent pink stain to the skulls of South American leaf-eared mice.

Most carnivorous rodents limit their attentions to insects. A few, such as *Rattus jalorensis* of Malaysia, have a preference for molluscs, and others sometimes tackle birds and mammals. The grasshopper mouse sometimes hunts and eats harvest mice, and many squirrels are active predators of vertebrates. In some parts of Canada, Franklin's ground squirrel has antagonised sportsmen by destroying large numbers of ducks' eggs and ducklings; in one study they were found to be as destructive as crows, destroying 19 per cent of the ducks' nests in their territory.[151] In America, the round-tailed ground squirrel has been observed to stalk English sparrows,[17] and thirteen-lined ground squirrels sometimes kill domestic chickens and young cottontail rabbits. Captive rats and mice frequently show cannibalistic tendencies, amicable partnerships of several years standing often ending with one partner being eaten by its associate. The discovery of a pet rodent, unconcernedly washing its face while sitting beside its headless mate, is unlikely to endear it to the owner, yet primitive human races often disposed of the old and ailing in a similar manner. As among human cannibals, the soft brain is invariably eaten first, probably because it requires no chewing. The cannibalistic rat may completely dispose of its victim's flesh and leave the skin intact and turned neatly inside out.

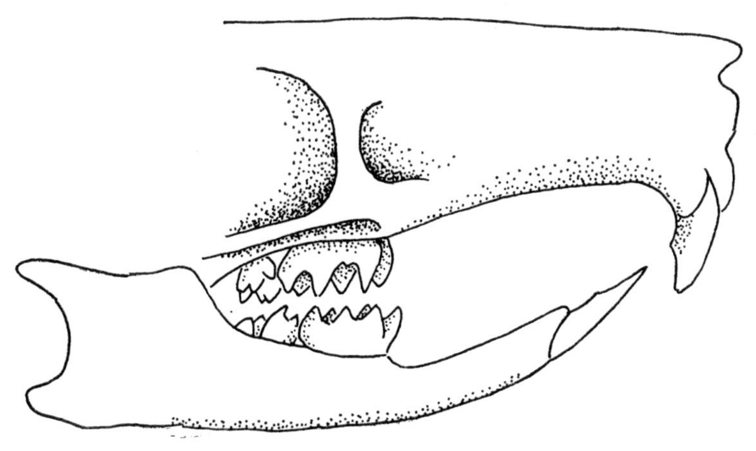

Fig 11 *Cheek-teeth of carnivorous rat,* Deomys

Many rodents which feed in exposed situations or are too large to rely on concealment, depend largely on their running ability, visual acuity and sense of hearing to escape predators. Speed has been attained by two different ways, one involving specialisation of both pairs of limbs, while the other affects only the hind limbs. Both types involve some structural modifications which parallel those which occurred in the evolution of hooved runners such as horses and deer. South American forms are particularly deer-like in general appearance through the elongation of fore and hind limbs and the best runners, the agouti and mara (page 54), also show a reduction of the hind toes to three on each foot, each hind toe bearing a hoof-like claw. Both animals are diurnal but live in quite different situations. The agouti lives in damp, well-vegetated regions, each in its own runway system where a short burst of speed enables it to reach cover. The mara, however, lives in the open pampa and stony wastes of Argentina and Patagonia. When alarmed, it does not go to earth but runs at high speed for a considerable distance. The mara's long front legs seem to be something of an embarrassment when the animal is resting; they are either fully extended like those of the Sphinx, or folded under the chest in a cat-like manner. Another feature of this species is the long eyelashes which give protection from the sun's glare. No work seems to have been carried out on how small diurnal rodents can tolerate excessive radiation from the sun, but in two genera of African grass mice (*Rhabdomys* and *Lemniscomys*) the braincase is covered with a black-pigmented membrane which possibly serves this purpose.

In other parts of the world, a number of surface rodents show enormous development of the hind legs. With the muscle power packed into two limbs instead of four, speed and manoeuvrability are increased. But as a small boy discovers after changing from pedal-car to bicycle, the slightest obstruction or fault with the steering can lead to a tumble; for this reason, bipedal species are most common in sandy deserts and areas with little vegetation. During the North African Campaign of 1941–2, many servicemen became acquainted with an African bipedal species, the jerboa or desert-rat, which came to be figured on military insignia. The hardy and graceful jerboas were described by Herodotus some 2,000 years ago; he called them 'dipodes', or 'two-footed', and the name lives on

Fig 12 *Desert jerboa,* Jaculus

Fig 13 *Hind foot of a desert jerboa*

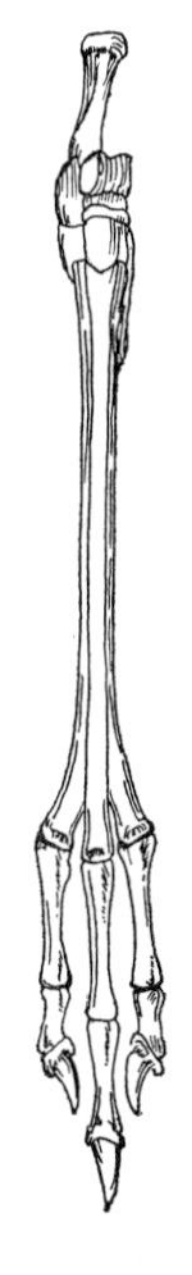

Fig 14 *Bones of hind foot of a desert jerboa*

as the family Dipodidae. There are twenty-five species of jerboa, only three living in Africa, the rest in Asia. They vary in size from a dwarf jerboa of the Gobi, *Salpingotus crassicauda*, which, with its body length of 41mm, is probably the world's smallest rodent, to the giant jerboa which is 200mm long. The hind legs of most species, besides being at least four times the length of the front pair, are further specialised in having the three middle foot-bones, the metatarsals, fused to form a single cannon bone. As in horses and deer, this gives great strength and support, lifting the animal off its feet and on to its toes when running. There is often a reduction of the hind toes to three; when the first and fifth toes are present, they are usually very small. Different species vary in the ways they run: some make double tracks in the sand while others make one in front of the other in a diagonal line. The large comb-toed jerboa, known only from Turkestan, inhabits areas of shifting sand, and its long hind toes are furnished with combs of bristly hairs which function like sand-shoes. The dwarf jerboas have especially large pads of springy bristles beneath the hind toes, giving each foot the appearance of a miniature toothbrush. Besides saving wear on the feet and minimising heat loss or gain, the pads probably contribute to the animal's jumping ability. Jerboas living in sandy areas have tufts of hair around the ear opening to keep out dust, and the African jerboa (*Jaculus*) has a thickened fold of skin which can be drawn down to protect the nostrils when the animal is digging.

The biology of *Jaculus jaculus* has been studied in the Sudan by David Happold, who found that several different kinds of locomotion are employed. All four limbs are used when the jerboa is hunting for food, but at other times it is bipedal. When just pottering about, it takes short hops on its hind legs. At medium speeds these limbs move independently in a series of strides, while at high speeds they seem to work together to give a maximum thrust. When alarmed, some jerboas can cover 3m in a single bound and clear grass over 500mm high. The rough-legged jerboa uses its leaping powers to get into bushes when ground vegetation is scarce; it leaps high in the air, grabs a twig with its teeth and front legs, then scrambles

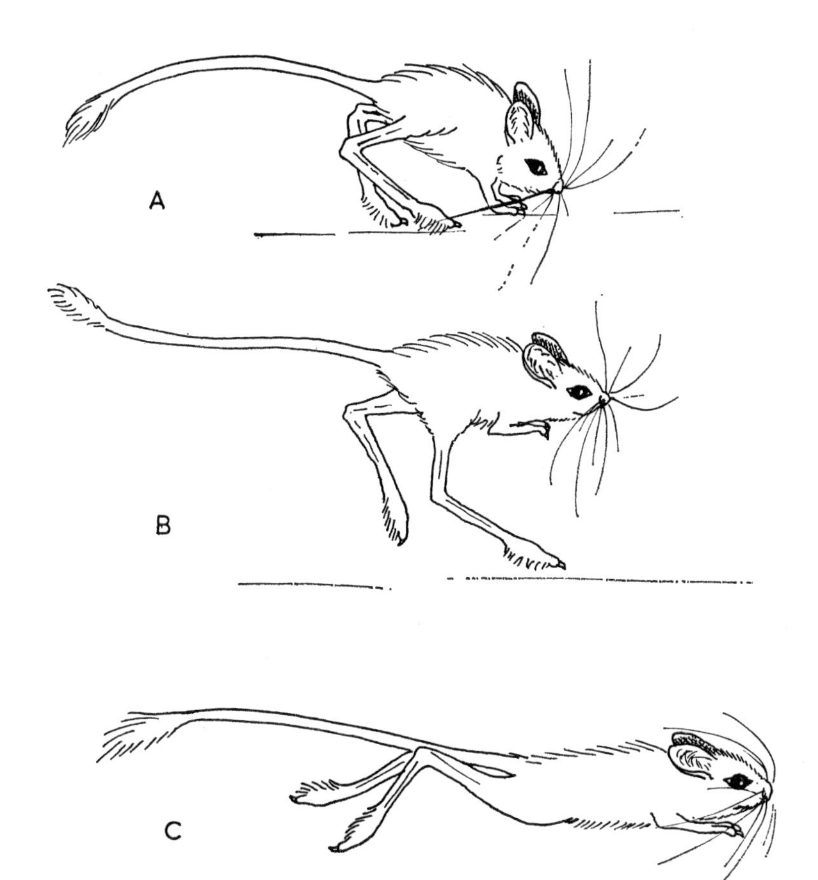

Fig 15 *Mode of progression of jerboa,* Jaculus jaculus*:*
(A) *movement on all fours;* (B) *medium speed;* (C) *fast speed*

into the crown of the bush which may be a metre or so above the ground. Any human athlete who performs feats like those of a jerboa is likely to suffer from 'slipped disc', through the continual jarring of the vertebral cartilages. It is not known whether jerboas ever experience the misery of lumbago, but at least they cannot suffer from fibrositis because the vertebrae of the neck region are short and expanded and many of them are fused together, with only the atlas always free. Some dwarf jerboas not only have six neck vertebrae fused, but also the first three of the chest. A long tail is essential to a bipedal rodent, serving as a prop when the animal is resting, and as counterpoise, rudder and brake during running; the tail streams out backwards on take-off and is raised on landing. The tails of most bipedal species are tipped with a tuft of conspicuous black or white hairs, but as yet, no experimenter has wielded a pair of scissors or paintbrush in an attempt to determine the function.

Bipedalism has been adopted by members of several families other than jerboas. The largest is the spring hare (page 72), common in many sandy regions in southern Africa. The name is appropriate as the animal's appearance suggests a cross between hare and kangaroo. With its long legs it can cover 2m in a single bound, and in periods of drought it may wander 30km a night in search of food and water. Among other unusual features of its anatomy, all five fingers on each foot are well developed with long claws, probably as an adaptation for digging. In Australia, ten species of hopping mice (*Notomys*) (page 71) live in lightly timbered, rather sandy country; and in North America there are two genera of jerboa-like rodents, the kangaroo mice (*Microdipodops*) and kangaroo rats (*Dipodomys*). All these forms show structural modifications similar to those of jerboas, kangaroo rats even having some of the neck vertebrae fused. Twenty-two species of kangaroo rats occur from south-west Canada to Mexico, while the two species of kangaroo mice are confined to Nevada and adjacent states. Besides their bipedal habit, members of both genera are well suited for arid environments—kangaroo mice have storage tissues in the tail like some jerboas. Not all bipedal species inhabit dry country. One group known as jumping mice live in moist habitats in North America, Europe and Asia and are included by some authorities with the jerboas. Others regard them as a separate family, the Zapodidae. Jumping mice resemble dwarf jerboas in having the central foot-bones free, not fused; but they are still able to jump 2m, relying on the tail as a balancer and turning a somersault if it is broken. American jumping mice live in wet meadows and woodland glades, sometimes by the banks of small streams; they are all good swimmers. The Old World members of the group include the birch mice of Europe and Asia, species which are more adapted for climbing than running.

In Asia and Africa, there are over eighty species of rat-like rodents which, although not bipedal, resemble jerboas and kangaroo rats in several respects. They are the jirds and gerbils, which are common in many dry habitats. A species which is familiar to many people, in all parts of the world, is the Mongolian gerbil, *Meriones unguiculatus*. This rodent's rate of spread, via pet stores and laboratories, rivals that of the hamster. In 1935, twenty pairs were captured near the Amur River in east Mongolia, and taken to Japan; from there, four breeding pairs were imported to the USA in 1954 for use as laboratory animals; ten years later, twelve pairs of American stock were taken to Britain. Gerbils thrive in captivity and seldom bite when handled. They are quite sociable animals, and being adapted to life in arid areas, are able to subsist entirely on dry food.

Rodents living in relatively barren areas may be unable to conceal themselves while feeding, but absence of ground-cover can be advantageous in allowing a wide field of vision and unimpeded hearing. Gerbils, jerboas and diurnal ground squirrels and marmots, habitually stand on their hind legs to scan the surroundings; bipedalism probably first evolved as an aid to this more effective use of sight.

Although nocturnal jerboas, kangaroo rats and gerbils have very large eyes, they rely even more on an acute sense of hearing to give warning of predators. Not only do they have large external ears; the bony capsules of the middle ear are much bigger than those of other rodents too. The early-warning system of the African gerbil, *Tatera leucogaster*, is highly effective so far as detecting owls is concerned.

In one locality where I trapped small rodents for two years, gerbils formed 9 per cent of my catch, but in the same period only 2·7 per cent of the rodent skulls in owl pellets belonged to this species. The ear capsules of the gerbil, large as they are, seem insignificant when compared with those of the kangaroo rats, kangaroo mice and dwarf jerboas (Fig 16). The significance of these enormous ear

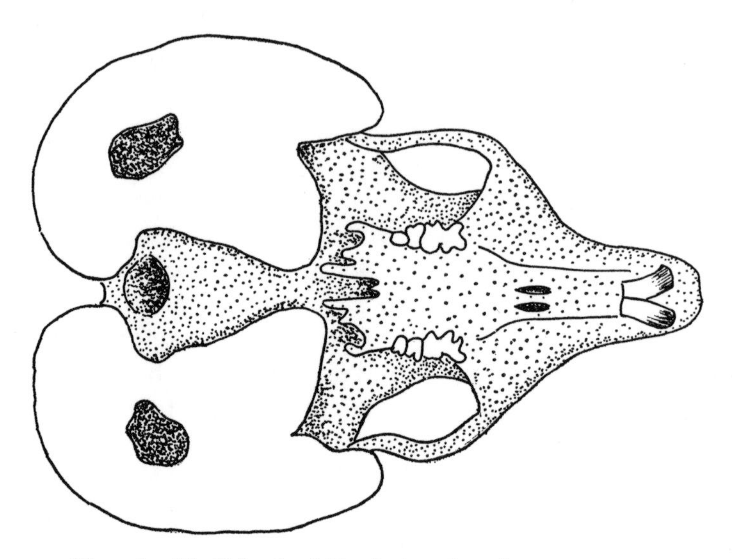

Fig 16 *Skull (underside) of American kangaroo mouse,* Microdipodops, *showing large auditory capsules*

capsules can only be appreciated with some understanding of the mechanics of hearing. The middle ear, whether of man or rodent, serves as a kind of transformer, changing airborn vibrations received by the eardrum to waterborn vibrations which can be sent along the fluid-filled inner ear to reach the nerve endings. It can be visualised as a box with a large, membrane-covered hole at one end which is linked by a lever system of small bones to a small membrane-covered hole at the other end—movement of the outer membrane, the eardrum, being amplified at the inner membrane.

Webster,[170] in his studies on the hearing of kangaroo rats, showed that the large middle-ear chamber serves two functions: besides allowing the lever system to be longer and consequently more powerful, it reduces the resistance to movement of the eardrum because the large volume of air is only slightly compressed. In the human ear, the movement of the eardrum is amplified about eighteen times, but in the ear of the kangaroo rat, amplification is ninety-two times, the sense of hearing being four times as acute. After finding that kangaroo rats were most sensitive to low frequency sounds of between 1,000 and 3,000 cycles per second, Webster took his tape recorder to the Arizona desert in an effort to find whether this kind of sound was made by natural predators such as snakes and owls, and whether the kangaroo rats could avoid being caught. By making his observations under red light which is invisible to nocturnal animals, Webster found that a flying owl made sounds of up to 1,200cps, sufficient for a rat to detect and leap away just before it could be seized. Rats were also able to avoid the sidewinder rattlesnake which produced brief bursts of 2,000cps just before striking. In every instance, the kangaroo rat leapt out of the way just an instant before the predator struck.

The reproductive period is a dangerous time for most kinds of rodents, both for the inexperienced young and the pregnant and lactating females whose increased food requirements necessitate more intensive foraging. Surface-dwellers which lack the protection of a burrow are, of course, especially vulnerable to predators at this time. In several species, natural selection has favoured the production of precocious young which, like those of deer, cattle and some ground-nesting birds, are physically well-developed at birth, or at least are able to live an independent existence in a very short time. The newly born agouti is covered with hair and has its eyes open, and within an hour it attempts to nibble vegetation. Gundis of North Africa and chinchillas of the Andes are able to run soon after being born. Even the American porcupine cannot afford a prolonged infantile stage, and a youngster has been observed to stand three minutes after birth, and use its tail as a weapon of defence only forty-five minutes later. The African harsh-furred rat also shows accelerated early development. In Uganda, M. J. Delany reared from birth the young of three species of rats collected from the same forest. He found that the two arboreal, burrowing species, *Praomys* and *Hylomyscus*, developed at the normal rate for most small burrowers and opened their eyes when about fourteen days old; the surface-living harsh-furred rats, however, opened their eyes on the fifth day and were fully-furred miniatures of the adult by the time they were nine days old.

In arid areas, a lactating female rodent faces increasing demands on her body fluids which may be difficult to replace. Some species of gerbil partly overcome the problem by reducing litter size and producing precocious young which are soon able to feed themselves. Among rodents living in northern regions or at high altitudes where breeding seasons have to be brief, there is a tendency for the females to produce a single large litter each year, so limiting the number of times the parent is at risk and also giving the young time to mature sufficiently before the end of the summer. The young of hibernating species must build up enough energy reserves to last them the winter, and at least in ground squirrels there is evidence that the young of hibernating species grow far more quickly than those of non-hibernators.

5

BURROWERS EXTRAORDINARY

The mole is so often selected by teachers and textbooks as an example of a burrowing mammal that there is a tendency to forget that its digging ability is rivalled by several species of rodents. Before making comparisons, it should be mentioned that the mole has a different identity in different continents. The name is applied to members of two families of insectivore: the Talpidae which occurs throughout the northern hemisphere, and the Chrysochloridae which are the golden moles of Africa. There is also a marsupial mole in Australia. Although they are not closely related, the two insectivorous families can be considered together since they are similar in appearance and habits. Both families have a fair record of proliferation, with nineteen species of talpid moles and fourteen species of golden moles. As for rodents, about thirty-eight species spend virtually their whole lives underground, with about another seventy rather less specialised, often coming up to feed on the surface. Representatives of no less than seven rodent families have become adapted to subterranean life. In distribution they are ahead of the burrowing insectivores, having penetrated to South America, but the Himalayas seem to have been a barrier to all burrowing mammals.

Burrowing rodents and burrowing insectivores seldom share the same habitat, but usually occur in quite different types of soil. Moles, which feed chiefly upon worms and insects, are confined to moist soil rich in humus. The normally vegetarian rodent is less restricted in dietary needs. Most burrowing species have made a niche in relatively infertile and often arid soils such as the prairies and savannas—even in steppes and semi-desert areas where the soil is sometimes full of rocks and stones. From a human viewpoint, it may seem strange that burrowers have evolved to cope with habitats which are not only difficult to dig but which provide little ground-cover apart from grass and herbaceous plants and where the surface may be barren for part of the year. But it must be remembered that although the soil surface may be bare, there is often a rich supply of potential food just below. Harsh climates and poor soils which are inadequate for trees, often encourage plants with tap roots and storage structures such as bulbs and tubers, which form the burrowing rodent's staple diet.

Both burrowing insectivores and rodents share a number of common features. They include small eyes and ears; soft, dense and usually short fur; well-developed shoulders and forelimbs; and feet which are specialised for digging. There the similarities end, and no one can fail to distinguish insectivore from rodent. The mole, with its long delicate snout and flabby body, looks like an underground animal, a precision instrument designed only for insinuating itself through the soil with minimum effort. The whole skeleton, muscular system and general body form is adapted for tunnelling. Even the penis is directed to the stern. On the surface, the mole is quite helpless, for the articulation of the lower arm forces the hand to be turned permanently outwards—an excellent position for digging and shuffling through tunnels, but almost useless for anything else. The animal is extremely vulnerable and can be killed by a light tap on the nose or even by a loud noise.

The burrowing rodent, however, gives the impression of being able to take care of itself in any situation, for it has a firm round body which is lifted well off the ground by sturdy limbs, and a head like a small cannon ball which is often armed with a pair of protruding and dangerous-looking incisors. On the surface, some species adopt an aggressive stance when threatened, facing their persecutor and grinding the teeth. Sometimes the animal advances with its jaws agape and attempts to bite. This apparent fearlessness probably stands the rodent in good stead when underground, since any ready-made burrow is attractive to rats, mice and snakes. Once, when I kept a Hottentot mole rat (*Cryptomys*) in a box of soil, a house mouse was accidentally introduced. A few hours later, the mouse was found with its head partially eaten. By the following morning the crushed body had been buried in the mole rat's hole. The occasionally carnivorous habits of this species had been noticed as long ago as 1911, when someone fed his pet mole rat upon earthworms.

The adaptive significance of the differences between moles and burrowing rodents can be appreciated when their habitats are compared. Working in soft soil, the mole runs little risk of damaging its snout; the head never seems to be used in digging and the hands are pushed forward alternatively, operating rather like snow-ploughs in front of the sensitive worm-locator. Incidentally, the marsupial mole shows less regard for this appendage, for it is protected by a horny shield. The rodent, on the other hand, must often dig through coarse gravel and be able to extract any large stones which bar the route. The hands, besides serving as efficient shovels, must be able to remove any obstacles encountered. Consequently the rodent's forelimbs are less specialised than those of the mole, and when digging, they usually operate in rather a dog-like fashion, below the body rather than at the side. Since the head has to be close to the working face of the excavation, it needs the protection of a tough rounded skull. The skull of the African mole rat (*Heliophobius*) (page 71) was found to be so strong that a steel rod, a square millimetre in section, only penetrated when a weight of 3kg was applied —this is equivalent to about 8cwt per sq in. The massive appearance of the head is due largely to the enormous incisors whose roots often extend the whole length of the jaw. Such powerful tools need well-developed muscles for their operation, and in many burrowing species with reduced eyes, the greater part of the sockets are taken over by part of the jaw musculature.

There is considerable variation in the burrowing methods employed. Some species rely chiefly on the strong claws of the forefeet. Others, which have projecting upper incisors, use these in digging and levering out roots and stones. The lower incisors of the African mole rat (*Cryptomys*) are able to move independently of one another to loosen the soil. If one of these animals is placed in a box of well-packed soil, it goes to work like a small grey mechanical shovel operating at full throttle. The jaws rhythmically open, slice into the earth, close, then fling the load to the side where it is kicked away by the feet. In a few minutes the animal disappears below the surface. When the incisors are employed for digging, the lateral parts of the upper lips close behind them to prevent soil entering the mouth.

In addition to the actual excavating, any burrower has to clear loose soil from the tunnel. Some move backwards and push it with the hind feet and tail, while the East African mole rats (*Tachyoryctes*) turn round and push the spoil-heap with one side of the face and one foot. Burrowing rodents seem to experience no difficulty in reversing in their narrow tunnels. *Cryptomys* simply pushes its head under its body until it is lying on its back, and then twists upright; pocket gophers and several species of mole rats probably avoid such contortions whenever possible since they can run backwards almost as fast as forwards.

Not all burrowing rodents are confined to rough and barren ground. Some species regularly emerge to feed on the surface, in spite of the risk from surface predators. Those few species which have attempted to enjoy the best of both worlds are all confined to tropical jungles where there is plenty of cover. The bamboo rats (*Rhizomys*) of south-east Asia dig extensive burrows under clumps of bamboo, the roots of which form their staple diet. At night, these rats frequently leave the safety of their burrows to raid plantations of tapioca and sugar cane, often being killed by road traffic on the journey. Although *Rhizomys* is one of the less specialised burrowers, retaining a tail up to half its body length, it has well-developed claws for digging. The presence of succulent bamboo stems may lead this rat to disregard the role for which it evolved. It climbs the bamboo by straddling the stems and gripping with its legs, then cuts regular sections with its vertical incisors and carries them back to the burrow. The related lesser bamboo rat of Nepal and Burma also leaves its burrow at night to forage for green leaves, in some areas having a predilection for tea. In the New World, the Brazilian shrew mouse (*Blarinomys*) also lives in forests, making its burrows under the leaf litter.

Even in exposed grasslands, some burrowers may crop surface vegetation. Those familiar North American burrowers, the pocket gophers, do not confine their attentions to underground roots. A study made in Colorado of the plains' pocket gopher, *Geomys bursarius*, showed that grasses formed its principal food except in spring. Although the limbs are specialised for digging, they can still carry out other functions. The plains' pocket gopher is extremely fastidious in cleaning its food, passing leaves through its hands for twenty or thirty seconds before eating. It is dexterous enough to manipulate kernels of rolled barley. Using a microscope, one American mammalogist examined stomach samples from nearly 300 pocket gophers and found that almost no soil had been ingested.[166] The Hottentot mole rat also holds food in its hands, frequently shaking them to remove soil. Before eating a sweet potato or bulb, it peels away a small part of the skin with its incisors, then the centre is eaten and the rest of the skin left intact.[38] Mole rats, pocket gophers and other specialised burrowers do not hibernate, but those living in severe climates store food in their burrows, either in special storage chambers or in the nest. Mole rats of Asia and Europe have been found with autumn stores weighing about 14kg. For one species, *Spalax monticolor*, stores up to 50kg have been recorded.[110]

In South America, south of the Amazon forests, there are vast tracks of grassland. The pampa is equivalent to the North American prairies, while in the plateaux regions of the Andes, known as the altiplano, there are bleak, grassy areas which experience sub-arctic conditions. The pocket gophers have not penetrated further south than Panama, but from southern Peru to Tierra del Fuego the grasslands are dominated by another, quite unrelated burrower, the tucu-tuco (Fig 18). There are about thirty species of tucu-tuco, members of the family Ctenomyidae, a name which refers to the combs of bristles on the hind feet. Apart from these combs, and of course the absence of cheek pouches, the tucu-tucos greatly resemble the pocket gophers, in appearance as well as habits. On sunny days the animal may be seen on the surface, seldom more than a metre from its burrow. Compared with other burrowing rodents, its eyesight is quite good, for it can distinguish a moving person 50m away. In the open pampa the greatest dangers are likely to come from above, but the tucu-tuco's eyes are situated high on its head and able to scan the skies from the safety of the burrow entrance. The curious name is derived from the call-notes uttered by some species, especially when an animal walks over the burrow system. The most picturesque description was in 1822 by the great naturalist W. H. Hudson. Referring to the tucu-tuco in the pampa of La Plata, Hudson wrote:

> There it is found living; not seen, but heard, for all day long and all night sounds its voice, resonant and loud, like a succession of blows from a hammer; as if a company of gnomes were toiling far down underfoot, beating on their anvils first with strong measured strokes, then with lighter and faster, and with a swing and a rhythm as if the little men were beating in time to some rude chant unheard above the surface.

The purpose of such vociferousness is still a mystery. Most burrowing rodents, including the tucu-tuco, have permanent burrow systems which serve both as a home and a feeding ground. Usually two types of tunnel can be recognised: deep tunnels serving as living quarters and containing the nest; and long, winding tunnels near the surface used only when searching for food. Just as a mining company may invest in a single main-shaft and dig out subsidiary tunnels for as long as the lode persists, so the rodent remains in its home burrow while there is sufficient food. Naturally the length of the feeding tunnels depend upon the age and sex of the tenant, as well as the abundance of roots and other food materials.

Depending on the animal's appetite and the rate of proliferation of food plants, a single burrow may supply the lifetime needs of its

Fig 17 (above) *Pocket gopher;* Fig 18 (below) *Tucu-tuco*

occupant, but if the plants are consumed before new seeds and bulbs can set, a permanent burrow could be a disastrous investment. The coruros (*Spalacopus*), a genus of two species which are confined to Chile, have overcome this problem by being nomadic. The coruro does not look like an animal which feeds underground, for it has a modest tail, external ears and large eyes; only the large, protruding incisors suggest a burrowing habit. Very little was known about these animals until Oswaldo Reig studied them in the coastal region of central Chile in 1966. He found that in this area the coruros fed almost entirely on the tubers and underground stems of huilli, a species of lily, and live in small colonies. The occurrence of the animals in an area can be recognised by the characteristic mound of excavated soil in front of each hole. These numerous holes are about a metre apart, and connect with each other and with feeding areas under clumps of huilli by a complex network of tunnels. When the lily crop in one area has been systematically undermined and eaten, the colony moves on to another. Reig considered an average colony to consist of about fifteen individuals, but excavations are made at an incredible speed; one colony was seen to move into a new area and dig over 250 holes in three days. Generally, a large number of colonies exist side by side, and their burrows interconnect. In one district, coruro workings extended continuously for 50km.

Burrowing rodents have reached their peak of specialisation in the Old World. Although there are nine genera belonging to four families, the term 'mole-rat' is popularly applied to most of them, two genera being known as mole lemmings. Mole rats are widely distributed in southern Europe, most of Asia and in Africa. A glance at the distribution map (Fig 19) shows that like their New World counterparts, mole rats are mainly confined to grassland, but some of the cricetid members also live in open woodlands. With the possible exception of the Eurasian mole rat (*Spalax*), the Old World burrowers spend most of their lives underground—not for them the half-way house of the pocket gophers and tucu-tucos. Mole rats have eyes which are no larger than pin-heads, and the ears are minute or absent; some species are blind and others virtually deaf. They con-

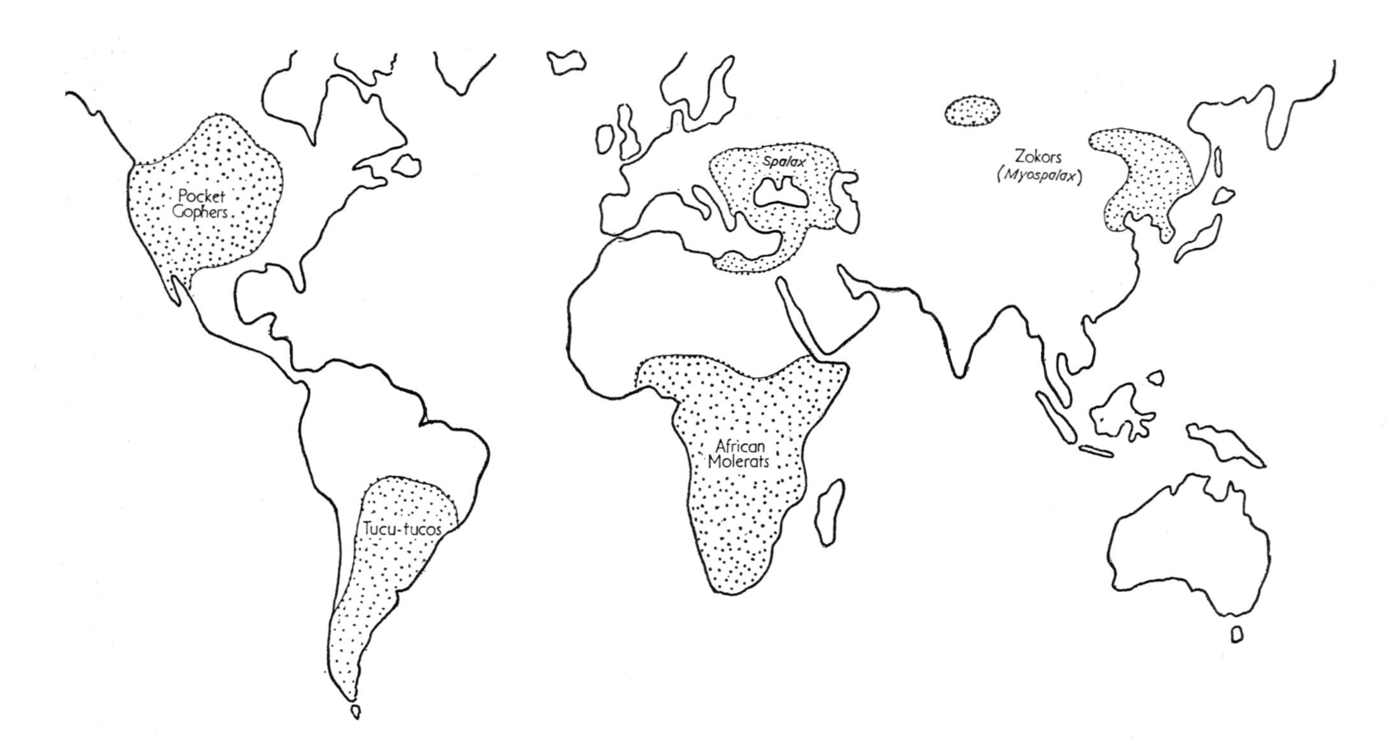

Fig 19 *The distribution of burrowing rodents*

tinually attempt to isolate their burrows from the outside world, and almost immediately seal any gaps made by an animal stumbling through the roof or by an inquisitive naturalist. Several observers, studying different species, have remarked how the mole rat senses a hole in its tunnel, possibly by a temperature change or by a draught. The African mole rat (*Cryptomys*) has bright, pin-head eyes but is totally blind and fails to respond to light or to the sound of the human voice. On the other hand, it is very sensitive to vibrations, also to air movements which are probably picked up by the vibrissae which are exceptionally fine and numerous; the finer hairs of the tail may also be sensory in function. They are certainly highly sensitive to touch. African farmers sometimes hunt mole rats for the pot. They simply find a recently made spoil-heap and scrape a hole in the roof of the adjacent tunnel; a few minutes later, if the occupant hasn't been alarmed, it pokes out the tip of its snout, giving the hunter a split second to strike down his spade; more often than not, the mole rat vanishes before its retreat is cut off. Rosevear, who spent many years in West Africa, described how the mole rat immediately becomes excited when a breach is made in its burrow; running towards it even when it is some distance away, periodically stopping to flatten its body and make pumping movements with the hind quarters. When the animal reaches the hole, it pokes its nose through before returning to collect loose soil for the repair; this is then dragged backwards and tamped into place by blows of the head.

In spite of their phobia against fresh air, the tunnel of every burrower must communicate with the outside world if excavated soil is to be removed. Since the exposed areas in which most of them live are infested by hawks and owls, mole rats are in the position of infantrymen in a frontline trench without a periscope, for the slightest exposure of the body may prove fatal. The zokors (*Myospalax*) of China and the USSR, instead of possessing enlarged incisors have enormous claws on the forefeet; they avoid undue exposure by pushing out earth with the outstretched forelegs. Several other genera compress the freshly dug, moist soil into plugs which are pressed up into the shaft to the surface; as fresh plugs are inserted from below, old ones are forced out of the top and a mound is built up; by this means, contact with the outside is avoided. The Hottentot mole rat, and probably other genera, compacts the earth plug by means of its hind feet. Irrespective of whether this animal is using its head for repairs or hind feet for plug-making, the tools concerned are vibrated at the astonishing rate of over twenty blows per second.[59] Species living in dry soil are unable to make plugs, and the removal of waste soil is a risky process unless it can be redistributed in abandoned tunnels. Usually the hind feet are used to kick the soil out of the shaft, and to an observer on the surface, it seems as if a miniature volcano is being formed; when not in use, the base of the crater is filled with soil.

Although a burrower may be oblivious of any dramatic temperature changes on the surface, it is extremely susceptible to variations in rainfall, and distribution probably depends as much on the moisture content of the soil as on its texture, or on the prevalence of food plants. The burrower has, of course, some room to manoeuvre since it can dig deeper when surface soils are too dry. Not all species are limited to well-drained soils, and the majority live in regions subjected to marked seasonal variations in rainfall; consequently there is often a noticeable change in habits during the wet and dry seasons. In Rhodesia, Richard Genelly noticed that the Hottentot mole rat frequently made its home burrow at the base of a termite hill where the soil was always moist, and the higher level prevented the food-store and nest chamber from becoming flooded. This species makes extensive tunnels when the soil is easily worked, but at other times it is confined to the home burrow. Members of a Eurasian family of mole rate, the Spalacidae, when living in areas which are liable to flooding, dig deep tunnels during the dry season but make large surface mounds for occupation during the rainy winter.

The Eurasian mole rats are more mole-like in general appearance than other burrowing rodents, having soft, nearly reversible fur, no external tail, and ears reduced to low ridges; they are unique among rodents in having no external openings for the minute eyes which lie just below the skin.

Like other burrowing forms, they feed mainly on bulbs and roots, but on occasion green plants are collected for the food stores. There is nothing remarkable in a blind animal leaving its burrow on dark nights when it has several advantages over sighted species, but one species of Eurasian mole rat has sometimes been seen wandering abroad in the daytime. Judging by the frequency its remains are found in owl pellets, forays outside the burrow must be a fairly regular occurrence. Possibly such suicidal missions are undertaken by a pregnant or lactating female to satisfy some inexplicable dietary craving, or perhaps by a victim of that irrepressible urge which afflicts all animals at some period in their lives.

Members of this family certainly hold the record for burrowing. The longest burrow recorded by the Soviet worker, Anisimov, had 114 mounds, the last two being 169m apart. Even the most enthusiastic inquirer could not be expected to dig up and measure a burrow system of these dimensions, but Anisimov did excavate the shortest burrow he found—363m long! This burrow represented 11cu m of soil; a two-month task for an animal which digs 0·2cu m a day, but has been known to extend its burrow by 2·5m per hour. Even with such speed, these mole rats are not shoddy workmen, and the chambers are kept in exemplary condition with walls free of roots and tightly tamped. Such tough and efficient burrowers do not waste energy unnecessarily. In winter, when the steppes are covered with snow, they leave the underground burrows and tunnel through the snow to dig out foodplants from above; when this becomes impossible, they retire below again and subsist on the winter stores.[57]

Very little is known about the breeding habits of mole rats. Nests with young have been found, but invariably the male parent was absent, and any association between the sexes is probably limited to one brief, blind encounter. In Israel, *Spalax ehrenbergi* has been extensively studied by Eviatar Nevo who found that just after the onset of the rainy winter season, this species builds special breeding mounds at the termini of certain burrows. The completed mounds may be over 2m in diameter and 1m in height; each contains a central nesting chamber surrounded by a complex of tunnels, storage chambers and latrines. A number of smaller mounds are scattered among the breeding mounds, and during the breeding season these are tenanted solely by males. During December and January, the nesting chamber in each large mound seems to serve as a honeymoon suite, suitors gaining admission through tunnels from their own mounds which connect discreetly with those of the bridal labyrinth. In February and March, the nesting chambers are tenanted only by a female, sometimes with her litter.

Not all mole rats restrict their social activities to brief conjugal interludes. There is one species (page 72), the naked mole rat, *Heterocephalus glaber*, which lives in small colonies of about a dozen individuals. Its distribution is limited to northern Kenya, Ethiopia and Somalia, but only in arid areas. The naked mole rat is about the size of a house mouse, but resembles a new-born rat in having a wrinkled, semi-transparent skin. It is not quite naked for there are a few long hairs widely scattered over the body and others form fringes round the feet which probably assist in burrowing. Apart from an East Indian bat, this rodent is the only land mammal without fur of some kind. For many years East African naturalists were puzzled by the speed at which these burrowers passed soil out of the holes. It seemed as if more than one animal was at work, yet the tunnels appeared to be too narrow for a pair of workers to pass. In 1971, Jarvis and Sale reported that members of a colony, besides sharing the same nest and food stores, also shared the digging; and not only is more than one mole rat involved, but a whole chain of workers form a kind of living conveyor belt. The animal at the earth face works away to loosen the soil, pushes it backwards, then, backing between the limbs of the animals behind, it kicks its spoil out of the hole; it then takes its place at the rear of the queue, shuffling towards the pit face as it straddles the workers coming back. As in human queues, there is sometimes impatience in reaching the front, and the relief sometimes forcibly pulls the digger from its work, often resulting in squeaked protests.

The lives and distribution of the burrowing rodents pose many unsolved problems. There is still no clue as to how they find their way in the long and intricate burrowing systems, which after all, are three-dimensional structures. Even more baffling is the means by which the animal can burrow through virgin soil to reach its goal. In 1951, Eloff discovered that if part of an African mole rat's tunnel is destroyed, the occupant digs another one parallel to make an accurate connection with the remaining intact portion. In experiments with two species, Eloff found that if the animal was placed on the surface, several metres from the nest chamber, it invariably burrowed directly to the chamber, no matter in which position it was placed. It is doubtful whether scent plays much part in homing for the sense of smell does not appear to be particularly well developed. Food seems to be recognised mainly by touch. When an African mole rat was presented with plaster balls covered with the integument from genuine bulbs, it immediately attacked them as food. This inability to discriminate between edible and inedible materials can lead to more than an unsatisfied appetite, for there have been reports of pets being accidentally poisoned after feeding on the toxic arum lily.

The burrowing rodents are all so specialised that their relationships with other groups are often obscure, but the burrowing habit has certainly evolved several times among quite different types of rodents inhabiting various parts of the world. As one would expect, the sociable types of burrower have not formed many species, but in the solitary forms, small populations have become isolated and genetic change has been rapid, leading to divergence into a large number of species and races. The fact that some species have colonised many thousands of square miles is not so remarkable, considering the rate at which some burrows can progress, and territorial expansion is not necessarily always underground. Dispersal may be speeded up in winter, at least in some regions. Pocket gophers, and mole rats of Europe and Asia, burrow in snow, and some of these snow tunnels have been reported to cross roads. Others, no doubt, cross frozen rivers and barriers which are otherwise impenetrable. Apparently inexplicable intervals of hundreds of miles between adjoining populations of some species may be due to climatic changes in the past making the intervening areas untenable; the zokor, for example, may have developed into distinct eastern and western forms when the central Mongolian desert cut across the middle of its range. In the south California desert, roads act as corridors for pocket gophers, enabling them to extend their range by keeping to narrow strips on each side. These strips are suitable for burrowing and feeding because of the run-off water they receive from the road.

6

LIVING UNDER DIFFICULTIES

Whether one lives in Miami or Timbuctoo, life presents problems. Some places, however, present more problems than others. The most difficult places to live in are the deserts—places we all know about but which are not easy to define. Most attempts have involved mean annual rainfall, and one of the most widely accepted definitions is that of Dr Koppen, who, in 1918, described deserts as regions with under 255mm of rain per year and a high temperature. Other authorities omit the high temperature qualification and include all the plant/animal communities at the dry end of the range.

Although important, rainfall figures are not an accurate guide to the aridity of a region. Rain falling in summer on the hot sands of the Sahara evaporates far more quickly than the winter rains of other parts of Africa. Temperature and relative humidity are also important factors. In many tropical, almost rainless zones, the temperatures fall at night and the relative humidity may reach saturation point and deposit dew—an important source of moisture for insects, snails and many kinds of succulent plants which spread their roots in complex networks just below the surface. There is even one species of mouse (*Leggadina*) of inland Western Australia, which constructs what appears to be a kind of dew trap. During the rainless part of the year, it piles a large number of uniform-sized pebbles on the ground above its burrow, forming a low mound about a metre in diameter. When the sun rises, the air in the spaces between the stones heats

more rapidly than the stones themselves; consequently dew is deposited on the pebbles, percolates to the burrow system below and ensures that it is kept cool and moist. The nature of the soil determines whether moisture is retained at a depth accessible to plants. Most desert perennials have very deep roots; the American mesquite, for example, can penetrate 15m in its search for moisture. It is not entirely then, lack of rain which produces a desert, but lack of biologically usable water. In the polar regions, the same condition applies, but here the amount of usable water is influenced by the extremely low temperatures. Although precipitation is often low and the cold air can carry little water vapour, immense quantities of surface water are locked in ice for three-quarters of the year.

To avoid difficult technical definitions, it is convenient to regard as desert all harsh, virtually treeless, sparcely peopled zones, irrespective of whether annual temperatures range from −2° C (28·4° F) to 52° C (125·2° F) as in North Africa, or from −56° C (−68·8° F) to 27° C (80·6° F) as in parts of Greenland. Apart from the continent of Antarctica where the only mammals are aquatic, these barren lands comprise a large proportion of the earth's land surface: hot deserts with an annual rainfall below 255mm cover about 20,000,000sq km, and the circumpolar tundra approximately 7,500,000sq km. If the steppe and alpine regions are included, the total area involved is about one-third of the earth's surface. In a world where space is at a premium, not only for the human species, the colonisation of this immense area, although difficult, is well worthwhile. In addition to these hostile realms, there are many other more temperate zones where harsh climatic conditions are experienced for only a limited part of the year. In these areas, many species of rodent have adopted similar survival procedures to the true desert-dweller.

The three main problems facing the desert-dweller concern shortage of water, lack of food and extremes of temperature. To a mammal of moderate size such as man, external temperatures present fewer problems than to a small rodent. This is because the rate of heat exchange between the body and the environment is directly related to the area of exposed surface as compared to size. As any housewife knows, small cakes are quicker to bake than a large one, although the same amounts of flour and heat are used. In spite of this additional handicap, rodents manage to live in places which man has only recently penetrated after epic feats of endurance.

Rodents living in arid zones, whether hot or cold, often show similarities in the ways in which they cope with difficult conditions. Just as the eskimo and bedouin rely on clothing for insulation, rodents which share the same regions have fine silky fur to restrict heat exchange and to reduce evaporation. The eskimo has a physiological adaptation in the layers of fat which protect the face from the intense cold; the hottentot of the Kalahari also has extensive fat deposits, but in the buttocks. Although the hottentot's fat may give its owner some comfort when sitting on sun-baked rocks, it serves primarily as a source of food and water in times of drought. Some desert rodents such as the kangaroo mice of Utah and California have fat reserves in the tail. The fat-tailed gerbil of North Africa (*Pachyuromys*) has a tail which is so full of fat that it resembles a small sausage. The beaver is similarly endowed and its tail shows a seasonal variation in a fat content, from 15 per cent in May to over 40 per cent in December, the tail swelling or shrinking according to whether its owner is lean or obese.[1]

In hot deserts, large mammals can combat the heat by evaporating water, but a small rodent cannot sweat or pant since its surface area would lose too much fluid. In one respect, the rodent's small size can be turned to advantage, since it can avoid extreme temperatures by retiring to its burrow. The majority of desert rodents seldom emerge from their underground homes during the heat of the day. In some areas, temperatures may vary by as much as 38° C in 24hr, but even a shallow burrow gives remarkable protection from such fluctuations. In Arizona, the temperature of the soil surface varies by over 80° C during the year, but at a depth of one metre the variation is only about 12° C, and the highest temperatures rarely exceed 30° C (86° F). Several species of hot-desert rodents can survive if caught in the open during the day, a few are even diurnal.

Those of us who have sunbathed unwisely at one time or another, may wonder how such species avoid heatstroke, especially as they cannot sweat to cool themselves down. A great deal is known about thermo-regulation in man—how the critical temperature is maintained by sweating, metabolism, shivering and so on—but work has only just begun on the temperature systems of other mammals.

The human body and brain, like a highly tuned engine, can only operate when the thermostat is critically set, but from the work of Schmidt-Nielsen and others, it seems that several species of rodents can live quite normally with fairly inefficient thermostats. The antelope ground squirrel, *Citellus leucurus*, of North America is active in the daytime when the temperature of the soil surface is as high as 65° C (149° F). The squirrel's body functions well at internal temperatures fluctuating between 32° C (88·8° F) and 43° C (109·4° F), but when the upper limit is approached, the animal either finds a cooler spot or runs into its burrow for a few minutes; there it flattens itself against the cooler surface and in 3 minutes can shed 4° C of body heat.[9] Some desert species employ special emergency measures when the body cannot be cooled sufficiently by other means, drooling copiously and smearing the saliva over the head with the forepaws. Obviously this method can only be used under extreme conditions and for short periods, or dehydration would result. Some species can tolerate dehydration to some extent: the antelope ground squirrel can lose water equivalent to a tenth of its body weight apparently without ill effects.

The importance of heat exchange by the body surface has already been mentioned, but this of course, is affected by the size of such structures as the ears, feet and tail. Throughout the mammals it is generally true that these structures are larger in animals living in hot areas than those of cold climates; the generalisation is often known as Allen's Rule. It may appear strange that a small desert rodent needs to increase its already large surface, but the body heat produced by metabolism must be dissipated with minimum loss of water; ears and tails with a scant covering of fur, act like the flutes of a radiator or air-cooled engine, and lose heat when the animal is in a cool place. It is rather strange that Micky Mouse, the epitome of a desert rodent, had his origins in Chicago. Rodents, seemingly of Mickey's kin, with protruding ears, big feet, bulging eyes and long tails (but without boots and pants) can be found in hot deserts the world over. Although so similar in general features, the majority are unrelated and their similarities are the result of convergent evolution.

Little research has been done on the function of tails and ears as heat regulators and it is important to remember that this function is only secondary; the short-tailed voles and lemmings, however, are characteristic of cold climates, while rats and mice predominate in warmer regions. The tails at least of the white rat and white-footed deermouse, are, in a sense, cold-blooded, for they normally take the temperature of the environment. When the animals are overheated or exercising, more blood flows through the tail and dissipates surplus heat. As further support for Allen's Rule, it has been discovered that white-footed deermice raised at 25° C (77° F) have longer tails than those reared at 15° C (59° F). This shows how easily geographical races can evolve.

At night, when the sand cools and the size of the appendages could become an embarrassment, some desert rodents simply fold the ears and sit on the tail. The largest ears in the rodent world are carried by the African spring hare (page 72) which conserves moisture and warmth by sleeping on its haunches with its head tucked between the thighs, and its long, furry tail curled about the head and body. The ears of this species are furnished with a flap of skin called a tragus which can be folded back to close the ear opening. The tragus is probably more effective than a burnous in keeping the ears free from sand.

It is not difficult to find exceptions to Allen's Rule, for the tails of the beaver and muskrat, inhabitants of cold northern waters, are larger than those of most tropical species. These tails, so useful in swimming, can present problems in heat conservation. The fat-filled tail of the beaver is, however, conveniently flattened in shape, and can be tucked under the body in cold weather. In the USSR, beavers are reluctant to appear on the surface at temperatures below – 20° C (4° F) and individuals presumably caught out by a cold snap have tails shortened through frostbite. In less severe conditions, the tails of both beaver and muskrat serve as heat regulators, being permeated by networks of minute blood vessels which eliminate heat during activity in warm air. The beaver's tail has a control system at the base so its blood circulation can be closed down in cold conditions. The heat exchange mechanism of the muskrat's tail is under nervous control, for if the nerve supply to the tail is injured, the animal becomes hyperthermic.[76]

Many other rodents have long tails yet inhabit cold regions. Dormice and squirrels are familiar examples, but these have well-furred tails which can be wrapped round the body like a blanket. Mammals living in cold climates have denser, finer fur than their counterparts of warmer zones, and the most important fur-bearing rodents of commerce inhabit either northern regions or high plateaux. The fur of the chinchilla (page 89) is so dense that lice and fleas are unable to use it for a home. This species, incidentally, together with many of the other rodents of the high Andes, provide a further exception to Allen's Rule, for the ears are extremely large. This is due to the peculiar climatic conditions of arctic nights and sub-tropical days. Whereas hot-desert rodents are usually nocturnal, the inhabitants of the Andes are forced to spend the bitter nights in their burrows, and since they have such thick fur, need large ears for heat regulation during the day. Several cold-adapted species conserve heat and moisture by huddling together in family or social groups; a beaver's lodge usually houses several individuals, up to thirty bobak marmots may occupy the same winter burrow, and twenty American flying squirrels have been found sharing one shelter.

Many of the mechanisms which regulate body temperature, also serve to conserve water. The burrowing habit, and insulating effect of fur on the body and tail all minimise evaporation for, of course, water is essential for the metabolic activities of any animal. Some water must always be lost by evaporation from the lungs, in the urine and in the faeces, but in the desert rodent such losses are kept to a minimum by various physiological processes. For instance, it has been found that at a temperature of 40° C (104° F) the white rat loses by evaporation 15gm of water per kilo of body weight, but the Egyptian jerboa loses only 8gm; the faeces of the white rat also contain six times more water than those of the American kangaroo rat. Mammals, unlike reptiles and birds, are unable to excrete uric acid but desert rodents reduce water loss by excreting a very concentrated urine. The kidneys are powerful water-resorbing organs, the parts concerned being the loops of Henle, and there is a strong correlation between the length of these loops and the aridity of the habitat. The urine of most desert rodents is between three and five times more concentrated than that of man, the record for kidney efficiency going to the spiny mouse of Israel whose urine is six times more concentrated.[145] The popularity of gerbils and hamsters as pets is due in no small way to their desert adaptations of producing dry faeces and limited quantities of urine.

In spite of rigorous conservation, some water must always be lost. This can be made good by drinking, feeding on succulent plants containing water, or by metabolising dry food. Many species can maintain themselves on a diet of dry seeds or grain, for not only does the seed contain about 10 per cent of water by weight but additional water can be produced by the breakdown of the food in the body; the oxidation of 1gm of carbohydrate generates about 0·6gm of water. It is not always possible to increase water production by simply increasing metabolism, because the extra oxygen needed involves the lungs in more work and consequently more moisture may be lost by this route. High protein diets may be positively harmful to a rodent which needs to conserve water, since the breakdown of protein involves an increase in the excretion of urine. From the few studies carried out on this aspect, it appears that desert species have a low metabolic rate; the house mouse, which is well adapted to dry conditions, prefers to eat less and risk slow starvation rather than consume large amounts of dry protein and die from dehydration.[53]

The consumption of desert plants would seem to provide an easy solution to the rodent's problem in obtaining water. The majority are succulents, which, by means of complex root systems, manage to trap the occasional drops of dew or rain and store the precious fluid in fleshy leaves or stems. With water at such a premium in the desert, it is not surprising that various mechanisms have evolved to protect the plants from the attentions of thirsty animals. Many produce highly toxic substances such as oxalic acid, 5 gm of which may be lethal to man; in maritime areas, the tissues of some succulents contain salt concentrations which are higher than sea water. With such inhospitable plants as the only source of water, a thirsty traveller might well echo the cry of the Ancient Mariner, 'Water, water, everywhere, nor any drop to drink.' The desert rodents are not discouraged so easily, and many can tackle such plants with impunity, regardless of the toxins or amount of salt in the tissues. The golden hamster, which in its native Syrian desert probably fed on the autumn crocus (*Colchicum*), is very resistant to the poison colchicine; weight for weight, the hamster is over one hundred times more resistant than man. Several species can drink sea water without suffering ill-effects; they include some kangaroo rats of the south-west United States and the sand rat, *Psammomys obesus*, of North Africa. A few species can take even stronger drink; the antelope ground squirrel can drink water which is nearly one and a half times the concentration of sea water, while the house mouse can manage a concentration three times greater than sea water. In the East Californian desert, the saltbush, *Atriplex confertifolia*, is a common succulent but it has a high salt concentration in the surface tissues of the leaves. The chisel-toothed kangaroo rat, *Dipodomys microps*, has a remarkable method of exploiting this plant without putting a great strain on its kidneys. It climbs into the bushes, collects leaves and carries them to its burrow to be stored or eaten. Before a leaf is devoured, the rat holds it in the forefeet and draws each surface several times over the lower incisors to shave off the salty outer layers. After this treatment, the salt-free tissues remaining are eaten. This species is unique among kangaroo rats in having especially wide lower incisors which are able to carry out this task. The adaptation is invaluable since saltbush leaves are available throughout the year.[80]

The North American wood rat (*Neotoma*) is the gastronome par excellence of the rodent world. In the Arizona desert, the Opuntia cactus forms 90 per cent of its diet during the dry season. Anyone who has braved the Opuntia spines to pick one of the beautiful flowers will appreciate its challenge for the gourmet. In addition to the numerous needle-sharp spines, the plant is covered with minute silica needles or glochidia which break off and irritate the skin for hours afterwards. As if this protection was not enough, the tissues contain a fair quantity of oxalic acid. The wood rat is not only able to avoid impaling itself on the spines or perforating its mouth and stomach with glochidia, but it can actually metabolise the oxalic acid.[137]

Life in the desert is always difficult but at certain periods of the year food and water may be completely absent, and climatic conditions too rigorous for any kind of life on the surface. At this time, many rodents remain in their burrows and subsist upon food stores which were laid down under more favourable conditions. As early as AD 77, Pliny described how marmots subsisted on their stores during winter, but the Roman naturalist was exercising his imagination when he wrote,

> When the male or female is laden with grasse and herbs, as much as it can comprehend within all the foure legges, it lieth upon the back and with the said provision upon their bellies, and then cometh the other, and taketh hold by the tail with the mouth, and draweth the fellow into the earth . . . and hereupon it is, that all that time their backs are bare and the haire worn off.

Some species store food material throughout the year, but others only store prior to seasons of adversity. The latter cannot have any concept of approaching seasons and must respond to environmental cues such as temperature or light. North American flying squirrels, *Glaucomys volans*, were observed by Illar Muul to store nuts through the year. Captive flying squirrels stored about 20 nuts per night during summer and 300 per night in autumn; this was far more than necessary since an individual can live on 7 nuts a day. Muul demonstrated that a decrease in daylength accelerated exploratory and storage behaviour. In the wild, this period coincided with the ripening of nuts, and its timing ensured that nuts were not collected too soon, also that no energy was wasted in searching when they were all gone. The storage behaviour of laboratory rats differs according to sex. When deprived, both sexes store food and water-soaked balls of cotton wool, but the female continues to hoard when food and water are plentiful.[67]

Solitary species usually have single food stores which sometimes attain a considerable size, especially in those rodents which lay up stores throughout the year. One of the most compulsive hoarders is the golden hamster, whose name is derived from the German 'Hamstern' meaning 'hoard'. Considerable effort must go into building up the stores. The nest of an African pouched rat, *Beamys major*, contained 1,383 large seeds of the Mwabve tree, besides a quantity of grass seeds; the total store weighed 1·2kg. Since this species can only carry 8 seeds at a time, the store represented some 200 collecting trips. It would have been interesting to know whether the seeds were actually eaten, because at one time Mwabve seeds were widely used by native tribes as an ordeal poison.

Many of the more sociable rodents such as squirrels, practise scatter hoarding, having no single cache but hiding food in various places. Professor Ewer[49] considered the hoarding behaviour of her captive African ground squirrels, *Xerus erythropus*, to consist of six distinct actions. First, an animal picked up the seeds to be stored, searched for a suitable hiding-place and dug a hole; it then pressed the seeds into the hole with a few blows of the incisors, pushed back the earth with its forepaws, and finally camouflaged the spot with a dead leaf or stone. Similar behaviour has been reported in the American flying squirrel which uses the incisors to hammer nuts into tree crevices and soft soil, also in the South American green acouchi which digs small pits. Both species attempt to conceal the storage sites with surface litter. Ewer suggested that scatter hoarding may have evolved through replete individuals trying to hide food from their fellows. It ensures that an animal will pick up and eat food lying on the surface before hunting for stored food, just as a man may use the small change in his pockets in preference to opening his wallet.

Nuts and seeds are the foods which are most frequently stored, for they are ideal for the purpose, containing concentrated nourishment in small, easily handled packs. Several species store other kinds of food materials; beavers collect quantities of branches, and several species of vole cut grass and allow it to dry before storing as hay for the winter. The prize for husbandry should belong to an Asiatic species, *Alticola strelzovi*, which collects hay in the spring and autumn, storing it in niches in the rocks. Each family collects about 5kg. To prevent the hay being blown away by the wind, the voles close the entrance to the family hayloft by making a wall of stones. The walls are carefully maintained, and during repair work a single individual may collect 2kg of pebbles in a day.[57]

Several groups of rodents possess cheek pouches which reduce the number of collecting trips necessary for building up the larder. These structures serve as excellent collecting bags; they are always to hand, leave the paws free, and are of the expanding type. One of the first to be impressed by rodent cheek pouches was Sir Francis Drake, who, in his account of a voyage to 'The backside of America' (probably Chile), wrote of a certain species, 'Under her chinne is on either side a bag, into which she gathereth her meate, when she hath filled her bellie abroad.' The most common type of pouch, like those of the hamster, are extensions of the cheek cavities which run backwards beneath the skin to the shoulder, and are lined by the mucous membrane of the mouth. The kangaroo rats, pocket gophers and spiny pocket mouse of North America have a different type of pouch, formed by infoldings of skin and consequently lined with fur. Useful features of these pouches could be followed by bag manufacturers—they can be turned inside-out for cleaning and have no seams to collect debris. The kangaroo rat, like the squirrel, uses its front paws to pick up seeds; it crams the pouches full then hops back to the store; there it swipes its paws against the sides of its face and shoots out the seeds with the rapidity of a machine-gun.[170] Only dried seeds, of course, can be stored indefinitely without becoming mouldy. When a kangaroo rat collects fresh seeds, it does not always take them straight to the burrow but first buries them in a shallow pit in which they are dried by the sun. In due course the preserved seeds are carried down to the larder where they can be stored until required.

Sleep provides a useful, and not unpleasant means of passing through a difficult period, and it greatly reduces metabolic needs. Kirmiz, who studied Egyptian jerboas in the laboratory, found that the basal metabolism of these animals is 68 per cent that of the white rat. This low rate is due, at least partially, to the jerboas becoming lethargic at high environmental temperatures. Temperatures above 40° C (104° F) kills white rats but only causes sleeping jerboas to salivate.

Most of us know that awful feeling after being aroused from a deep sleep, we only recover after stretching our limbs or drinking a cup of coffee. During sleep our body thermostat may drop a notch or two, for there is no point in keeping the heating at full blast and wasting fuel. Many mammals can close the dampers even more, and for much longer than a few hours at a time. This is what happens during hibernation, when the body temperature does not merely fall a few degrees, but usually to just above the air temperature. For example, the body temperature of an hibernating ground squirrel (*Citellus*) is about 5° C (41° F) or 30° C below its normal temperature. Naturally all body activities must slow down, including the heart beat and the rate of respiration. In this condition the animal feels cold to the touch and often does not awake when handled.

Hibernation poses many problems. How is it timed? How can hibernators fast for long periods? What prevents them freezing to death if air temperatures fall too low? And how do they know when to wake up? The stimulus for hibernation varies in different species. In some, it seems to be associated with shortage of food, while others may go into hibernation following a fall in air temperature or reduction in daylength. At least one species of ground squirrel, however, has been known to hibernate in a room with constant temperature and continuous illumination.

One feature common to all hibernators is that at a certain time of the year, they overeat and become extremely fat. Fat deposits may amount to one-seventh of the total body weight of the American woodchuck, and half the weight of the edible dormouse. Apparently, prior to hibernation, certain satiety centres of the brain become blocked so the animal continues to eat when its stomach is full. This does not mean that the prospective hibernator rushes about searching for food; on the contrary, there is an inhibition of activity so no energy is wasted. It has been suggested that the state of the fat deposits decide both the time an animal will go into hibernation, and when it will emerge.[104] The fatty tissues are usually brown in colour, due to respiratory pigments which help in the oxidation of fatty acids. Although the temperature of the body surface may fluctuate during hibernation, the temperature of the brain remains constant, and the part which acts as the thermostat is always active. Thus, if anything untoward occurs, such as a severe cold spell, the thermostat apparently sets off a chain of events which result in the burning of the tissues known as brown fat. When this happens, the body warms very quickly; the ground squirrel being able to raise its temperature by 30° C in three hours. Another mystery of hibernation is how the animal avoids dehydration, for water must still be liberated, although at a slow rate. At one time it was believed that the metabolism of fat reserves liberated sufficient water, but it has since been shown that the oxidation of 100g fat involves a loss of 36g water. On the other hand, the breakdown of 100g skeletal muscle produces a gain of 37g water. Following laboratory experiments depriving white rats and ground squirrels (*Spermophilus*) of water, the rats lost their fat and the ground squirrels muscular tissue.[13] The ever-active brain also seems to reduce the hibernator's production of urine by causing the pituitary gland to produce anti-diuretic hormone.

With hibernation such a useful means of beating the cold, why can it not be used to combat the hot, dry seasons of other parts of the world? Many rodents in fact, do just this, but through a vagary of the etymologist, 'summer-sleepers' are said to aestivate. Aestivation may be induced by food shortage, or, unlike hibernation, by a rise in temperature; almost certainly, the state of the fat reserves also play an important part. Times of onset can be extremely regular; an Iranian ground squirrel, *Citellus fulvus*, always goes into aestivation between 5 and 15 June when the vegetation becomes desiccated. There is no real difference between aestivation and hibernation apart from the season when the 'sleep' occurs. The same species may do different things in various parts of its range. In the Volga Urals, for example, the northern three-toed jerboa hibernates, while in Turkmenia it is active through the year; in Africa, jerboas often aestivate.

A few species demonstrate conditions which are neither complete hibernation nor aestivation. Instead of going into a long 'sleep' during the most adverse season, they become torpid for a few hours each day. This kind of behaviour is known in other animals—bats become torpid during daytime, and hummingbirds at night. The first time I appreciated this phenomenon was while working with fat mice (*Steatomys*) in Africa. These mice, incidentally, have a most appropriate name, for their bodies are so full of fat they resemble little furry balls. Fat storage as a method of survival has rebounded to some extent so far as the fat mice are concerned. They are regarded as a succulent delicacy by many African tribes who hunt them with great tenacity; when captured, the mice are skewered and fried in their own fat. A captive fat mouse was once kept without food or water for thirty-six days; at the end of that time it had lost a third of its weight but appeared quite healthy. During the dry season, some captives spent the day in such a deep state of torpor that they could be roughly handled without waking (page 89). The body temperature was a couple of degrees above room temperature and the respiration was most irregular, several short pants being followed by a pause of up to three minutes. Just before dusk the mice woke up of their own accord and respired normally. In this case the torpid state was not induced by shortage of food or abnormal temperatures.

The forest dormouse of southern Asia and Europe also undergoes periods of torpidity during the day; this species has been recorded as having pauses of up to seventeen minutes between breaths. There is also a record of a leaf-eared mouse of the Peruvian desert which became torpid under severe conditions.

So far, we have covered the adaptations employed by rodents of the hot deserts, the steppes, and some of the less hostile zones where climatic problems are only acute for part of the year. What of the rodents of the circumpolar desert—the tundra? This zone is the most difficult for terrestrial animals to colonise. The muskrat and the beaver, which penetrate its fringes, are aquatic and so tend to avoid problems of intense cold, shortage of food and shortage of water. An additional difficulty imposed by high latitude is the shortage of daylight; for many months of the year the days are too brief for the growth of green plants, and in any case, give little time for foraging or hunting. One might imagine that a rodent could survive quite easily by laying up a food store for winter, and hibernating or sheltering in a burrow. This is prevented by the permafrost—a layer of rock-hard, perpetually frozen soil which lies just below the surface. In some places the permafrost is over 1·5km in thickness and rarely is it more than 30cm or so below the surface. There are a few areas such as the gravelly banks of rivers, where the ground squirrel, *Citellus undulatus*, is able to dig its burrow before the surface freezes; in winter, the burrow entrances are usually covered by between 150 and 180cm of snow.[58]

There is one group of rodents whose name is synonymous with icy wastes, and with suicide—the lemmings. Apart from man, the lemming (page 90) is ecologically the most important animal of the tundra. Birds of prey and carnivorous mammals such as the Arctic fox are almost entirely dependent upon lemmings for food. In winter, when their prey lives under the snow, these animals are forced to migrate considerable distances in search of other food. On the other hand, lemmings may destroy in a night some tenuous oasis of tundra plants which has taken a thousand years to establish. In summer, they sometimes dig a prodigious number of burrows, estimates of up to nearly 1,000 per 1,000sq m having been made. This turning of the soil allows the sun's heat to penetrate deeper and so reduce the depth of the permafrost. The lemming can both be blamed for destroying the tundra, and praised for reducing the polar desert.

Two genera of lemmings inhabit the northlands, the true lemmings (*Lemmus*), and the collared or varying lemmings (*Dicrostonyx*) which turn white in winter. At first sight, these conquerors of the frozen north are unimpressive, rather like hamsters in general appearance. A close examination reveals that the lemming is the converse of the tropical desert rodent. It has very small ears and feet, a minute tail and dense fur. The animal looks rather like a half-empty furry bag; when picked up by the scruff of its neck, the whole body seems to fall to the bottom of the bag. The loose skin gives valuable insulation and compares with the eskimo's loose style of clothing. It enables the lemming to flatten itself for concealment or to absorb the maximum amount of heat from its surroundings; its wearer can also turn completely round in a burrow which is no wider than its body. When the weather is very cold, the lemming can erect its fur and snuggle down into its own skin. To any dweller or traveller in the Arctic, the eyes and ears are the most difficult organs to protect from the extreme cold. The lemming has small eyes and nostrils which are well protected by fur, similarly, the long hairs just in front of the ears can be pressed back to cover the orifices. The bony ear capsule itself is very thick-walled.

In winter, the lemming cannot dig into the frozen soil but lives beneath the snow, making burrows some 15m long, with shorter side tunnels on either side. A round nest of grass is made in the main tunnel, and here there is no problem in keeping warm, at least to lemming standards, provided the snow is at least a metre deep. Snow is a poor conductor of heat, and at this depth the temperature can be 22° C warmer than on the surface. The very thick walls of the winter nests of the Norwegian lemming increase insulation, and the collared lemming often closes the entrance passage with snow for about a metre in order to prevent draughts.

The collared lemmings and Ob lemmings, which inhabit the coldest regions, breed in both winter and summer. Winter can present grave problems of food shortage, while in summer the young may face even colder nest temperatures because at this time the cold does not come from above, but from the permafrost below. The summer burrows of the collared lemming are no deeper than 20cm, and the temperature often a mere 4° C (39·2° F). Several individuals sometimes share the same nest to keep warm. The collared lemming is also known as the hoofed lemming on account of the curious horny shields which, in winter, develop on the third and fourth claws of the forefeet. In areas where 'the hard tundra snow scarcely yields to the blows of a heavy iron spade', the large winter claws undoubtedly help in digging, and unprotected fingers doing this work would lose excessive amounts of heat and face the danger of frostbite (Fig 20).

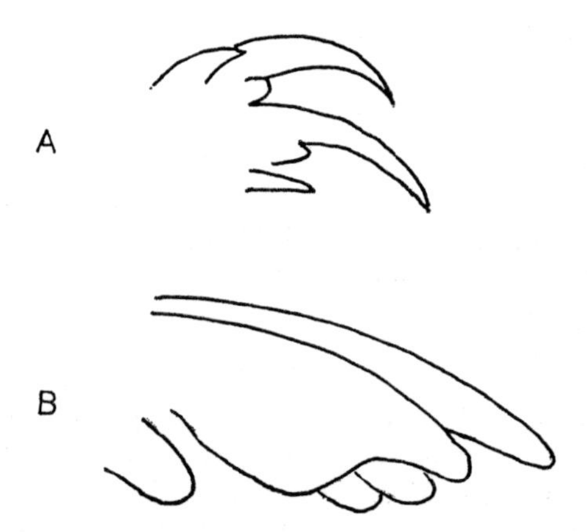

Fig 20 *Foreclaws of collared lemming,* Dicrostonyx*:* (A) *June;* (B) *October*

Lemmings neither hibernate nor store food, but must remain active to maintain body heat. Like some of the rodents of hot deserts, the lemming's body thermostat seems to have a wide range

Page 107 (*above*) African pigmy mouse, *Mus minutoides*—one of the world's smallest mammals. A monkey nut is at the left of the picture; (*below*) African striped rat, *Rhabdomys*

Page 108 (*above*) Striped hamster; (*below*) spiny mouse, *Acomys*

of tolerance; the body temperature of the Ob lemming (named after a Siberian river), fluctuates between 31° C (87·8° F) and 38·5° C (101·3° F). On protracted cooling it may even fall to 22·5° C (72·5° F) but below this point the animal succumbs. The food problem in winter is not so acute as might be imagined. Soviet workers, to whom we owe so much of our knowledge of Arctic animals, have discovered that 25 per cent of the leaves of grass and sedge remain green in spite of the lack of sunlight; not that the lemming is a food fad, for on the Kola Peninsular it was found that seventy different species of plants were eaten, including leaves, stems, berries, rhizomes and toadstools.

High winds are one of the greatest climatic dangers the lemmings may face during winter. No physiological adaptation can cope with such low air temperatures, and if the protective blanket of snow is swept away the little animals rapidly freeze to death. Of all the rodents in the world, there can be little doubt that the lemming occupies the most difficult habitat of all; by comparison even the spiny mice which feed entirely on bat droppings in the Aswan tombs lead rich and easy lives.

7

THE COSMOPOLITANS

Among man's less commendable characteristics are his wasteful and untidy feeding habits, and his inability to live happily unless surrounded by a variety of impedimenta. Considering that the human home, whether cave or penthouse, offers not only shelter but a host of hiding-places and plenty of food, it is surprising that only a few insects and rodents have attempted to share its comforts. Probably, this is because permanent human settlements are a comparatively recent innovation. Also, so far as rodents are concerned, this habitat is open only to those of small size which are able to utilise the same foods as man. There are, of course, many species which occasionally enter a human dwelling to enjoy its benefits for a time. But only three have been adaptable enough to accompany man far beyond their normal ranges, and to establish themselves eventually throughout the world.

The house mouse, *Mus musculus* (page 90), seems to have been the first rodent to become intimately associated with man. It is believed to have evolved from a wild subspecies, *Mus musculus wagneri*, which is still widely distributed over the dry steppes of central Asia, between the Caspian Sea and the Himalayas. Some of the earliest agricultural settlements have been excavated in this area, and it is thought that *Mus musculus* happened to be here at the right time to take advantage of man's new agricultural techniques and more stable way of life. Since the earliest times, its association with man was not without risk. In Afghanistan, pottery mousetraps of ingenious design have been recovered by archaeological excavations at sites dated 4000 BC. The house mouse is well adapted for living in human dwellings. It is small, nocturnal, extremely agile and alert. When ancient civilisations began trading with each other, many small animals must have been unwittingly carried among the baggage of the caravans crossing the desert, just as they are today in the holds of ships and aircraft. Most stowaways would soon perish on a rigorous journey of this kind, but the house mouse, provided it has access to grain and is able to make some kind of nest to maintain a high relative humidity, can survive for long periods without drinking. It can live for months on a diet of dry seeds, and its body can withstand considerable dehydration. The ability to survive is not the sole requirement of the successful colonist. A move into unfamiliar territory is dangerous for any animal, and a high reproductive rate is necessary to replace the unavoidable casualties. House mice face no disadvantage in this respect. Females begin breeding when about forty days old, have a gestation period of nineteen days and the litter size averages between five and seven.

The house mouse spread to the Middle East, the shores of the Mediterranean, then northwards through the whole of Europe. As the mouse populations became adapted to the various localities in which they found themselves, they diverged in form and habits from the ancestral stock, and a number of races, or sub-species, evolved. After Britain, France and Holland founded their American colonies, the northern race of the house mouse, *Mus musculus domesticus*, used the ships of these nations to establish itself in the northern half of the continent. Similarly, the southern race *M. m. brevirostris* was carried to Latin America in Spanish and Portuguese vessels. *Brevirostris* also became established in California and southern states, and hybrids occur where the two races meet. Other races of house mouse occur in Greenland, Alaska, India, Africa, Australia, and numerous oceanic islands. Where climatic conditions are mild, the house mouse may live out of doors but elsewhere it is generally an indoor animal; in some regions there are both wild and domestic populations. It has been less successful in the colonisation of Africa, possibly due to the absence of permanent sources of grain in some areas, and to competition from the multimammate rat which is a well-established domestic species. In parts of South America, it faces competition from laucha mice (*Calomys*). Some species of lauchas, and young multimammate rats, are so similar to house mice that they have sometimes been identified as such by zoologists. However, only mice of the genus *Mus* have notched incisors.

The house mouse is so adaptable that it can live in cold stores kept at temperatures around − 10° C (14° F), feeding entirely on frozen meat. More remarkably, such animals reproduce throughout the year, although fertility is reduced and there is heavy mortality among the young. Researchers at Glasgow reared fourteen generations of laboratory mice at − 3° C (26·6° F). They discovered that the first and second generations had shorter tails and heavier hearts and stomachs than their parents, no doubt reflecting a higher metabolic rate; they also had more fur, especially the females. The investigators were surprised to find that succeeding generations showed none of these abnormalities and were barely distinguishable from control mice reared in the warm; apparently they had become so physiologically adapted to cold that anatomical specialisation was unnecessary.[8] Although the house mouse is of minor economic importance when compared with rats, one individual can cause enormous damage in a short time, just by gnawing books and upholstery. Mouse plagues can have serious consequences. One of the most severe plagues in the present century took place in southern Australia in 1916–17, after an exceptionally good harvest. Grain stores valued at over £1,000,000 were completely ruined; one farmer put down poisoned bait and next morning found 28,000 dead mice on his veranda; at a wheat-yard, 70,000 were killed in one afternoon. In 1926, during a plague in California, a density of 25,000 mice per 1,000sq m was recorded.

The next rodent to achieve a world-wide distribution through its attachment to man, was the black rat, *Rattus rattus*, sometimes called the ship or roof rat. Historians, with so much human folly and intrigue to fill their pages, cannot be expected to record, or even notice, the movements of most wildlife, but the black rat marked its own indelible trail through world history. It is the major reservoir of human plague, and historical references to the disease give some idea of the animal's movements. The original home of the black rat was in India, a sub-continent isolated from the rest of the world by oceans and mountains, barriers which could only be penetrated by man. The first certain reference to bubonic plague was made in about AD 100 when Rufus of Ephesus mentioned 'deadly bubonic outbreaks in the Levant about 300 BC and in Libya about 50 BC'. This suggests that by that time the black rat had reached the eastern Mediterranean, probably as a result of trade between the Roman Empire and India. In AD 540, the great plague of Justinian erupted in the eastern part of the empire, reaching France in 547, possibly through rats carried in a ship from Egypt to Marseilles. Between the sixth and fourteenth centuries there is no authentic record of the disease in Europe, due perhaps, to the Saracen empire forming a barrier between India and the west.

The black rat was first recorded in the British Isles in 1187 when Gerald de Barry wrote that 'large mice, popularly called rats, have been expelled from the district of Ferns in Leinster by the curse of the Bishop Yvor whose books they had gnawed'. Plague first reached England in 1348, travelling from the port of Weymouth, via Oxford to London. There must have been a fair rat population by that time to account for the severity of the outbreak. The rat was certainly a familiar pest in Britain by the fourteenth century; Chaucer, in 'The Pardoner's Tale' mentioned that rat poison was sold by apothecaries. Official recognition of destruction caused by rats was shown by the Elizabethan 'Acte for preservation of Grayne', which included an authorisation to churchwardens to levy a tax in their parishes to provide a reward for vermin catchers; it set a price of one penny 'for the heades of everie three Rattes or twelve myse'. For several centuries the black rat was the major pest in all European countries and various control measures were tried. In 1489 the authorities of Frankfurt were so concerned that an attendant was stationed on the town bridge to pay a pfennig for every rat; after the tail was removed, the body was thrown into the river. Being predominantly a climbing species, and exceptionally unafraid of man, the black rat is eminently suited for travel on board ship. It probably reached the New World during the Elizabethan period, but since plague-infected rats are unlikely to survive a long journey, the date of its arrival is unknown; there is a record that Bermuda was invaded by black rats in 1615.[176]

Rats were introduced to many Pacific islands probably through the voyages of Captain Cook, who, in 1785, recorded that his ships were heavily infested. They were carried to other islands such as Bougainville by warships during World War II, and the colonisation of the most remote islets continues to this day. In the Pacific region, the black rat co-exists with the smaller Polynesian rat, *Rattus exulans*, itself a great traveller but confined to the Far East and the Pacific. With its greater climbing ability and omnivorous feeding habits, the black rat is able to live in a far greater variety of habitats than the longer-established species. It often makes its home in the thatched roofs of houses. Ten years after the last nuclear tests on Eniwetok Atoll, the black rat was found to be common on the most heavily devastated islands.[52] In Europe and America, it is generally confined to coastal areas, but it has penetrated deep into Africa. In several parts of the once 'Dark Continent', isolated pockets of black rats occur in remote areas, well inland; they probably mark staging posts on the old slave routes, or places of refuge in times of tribal wars. In addition to its economic significance in damaging food supplies, drains, cables, buildings, and all manner of commercial products from soap to plastics, the black rat can present a grave threat to indigenous wildlife. The flightless moorhen of Tristan da Cunha was exterminated by rats, and on Lord Howe Island in the Pacific, rats which came ashore from a shipwreck in 1918 destroyed four unique races of birds. The young of many species of sea bird are especially vulnerable, but the introduction of rats does not inevitably lead to their extermination; on an island off Tasmania, colonies of short-tailed shearwater suffer little or no mortality from the rat population.

The most important pest in present-day Europe and North America, is the brown rat, *Rattus norvegicus*. This species did not begin its wandering until a few centuries ago, and was unknown in the western world before the eighteenth century. It originated from temperate Asia in the region of the Caspian Sea. It is much larger than the black rat, and as would be expected in a species from a cold climate, the ears are short and covered with fine hair, not large and naked like those of its tropical relative; also the tail is much shorter. Apparently the brown rat was introduced into Copenhagen during a visit of the Russian fleet in 1716, and it is said to have reached England in 1728, Paris in 1750, Norway in 1762, the USA in 1775 and Spain in 1880. The species is a less agile climber than the black rat, but being a good burrower and exceptionally hardy, it is well adapted for the colonisation of colder countries. It soon became the dominant domestic species through most of Europe and North America, often living outside during summer and returning to the shelter of buildings in winter. It has managed to become established in the whaling station of South Georgia in the Antarctic, also in the Aleutians, where on some islands, it is the only rodent.

While the black rat usually lives in the roofs of buildings where its presence is often easily noticed, the brown rat prefers cellars and sewers, places where populations can increase without drawing attention until some flood, reconstruction work or other interference with their habitat drives them to the surface. In 1768, a rat-catcher described how, in some English houses, he caught black rats in the roof and brown rats in the cellar; when he caged the two species together, the brown rats promptly killed and ate the others. Nearly two centuries later, in 1949, investigators in London tried to check the relationship between the two species by keeping a colony of each in adjacent compartments and then removing the separating partition. The brown rats entered the neighbouring compartment almost immediately, leading the black rats to retreat to the ceiling. A few days later only the most aggressive of the male black rats were left alive, but none of the dead showed any sign of physical injury.[6] The replacement of the black rat by the brown is not entirely due to the more aggressive nature of the latter. Paradoxically, improved sanitation of cities by the introduction of underground sewage systems is favourable to the brown rat. On the other hand, the replacement of thatch by slate and tile, and of wattle and daub by brick and cement, reduce the number of habitats available for the black species. Although the brown rat is at some disadvantage in tropical countries, there is evidence that it is becoming well established in the sewage systems of some West African towns. Unlike the house mouse, domestic rats need plenty of drinking water and are unable to survive in arid regions, but even here, new territories have been opened for them by irrigation schemes.

Domestic rats are extremely difficult to trap or poison. The presence of an unfamiliar object can prevent a run being used for several days. Even when an animal becomes used to the presence of bait, it does not eat it all at once, but takes small samples at a time. This pattern of feeding gives the body time to absorb any poison, and if the animal becomes ill, it refuses to eat the same kind of food again. If, for example, the poisoned bait contained sugar, the rat may avoid any sugar mixtures which are put down subsequently. To counteract 'bait shyness', bait should be put down for some time before poison is added. DDT was once hailed as the ultimate weapon against insect pests. Subsequently the era of the rat seemed about to end when the Wisconsin Alumni Research Foundation gave its initials to a new anti-coagulant poison—Warfarin. This type of poison acts by reducing the clotting properties of the blood and eventually leads to a fatal haemorrhage. It is especially useful since it is slow-acting. If low concentrations are used, the victim suffers no illness for some time but continues to feed until a lethal dose has been consumed. The first set-back with the new weapon was noticed in 1958, when it failed to clear rats from a Scottish farm. Since then, brown rats resistant to anti-coagulants have been reported in several parts of England, Wales, and other European countries; in the summer of 1971, their presence was confirmed in a rural area near Raleigh, North Carolina. Resistant house mice were discovered in a food store at Harrogate, England, during 1960, and a little later, in Denmark and Germany. Resistance to poisons is not built up by individuals receiving sub-lethal dosages, but it is due to the occurrence of special genes for this characteristic in members of a population. It has been pointed out already, how the remarkable adaptive features of some rodents depend on quick genetic changes associated with a high breeding rate. When resistance genes are present in a population, the use of the poison does no more than select for resistant individuals. To overcome this danger, more than one type of poison should be used.

The occurrence of resistant populations of rats and mice is not the only problem faced by public health officials. In 1972, officials in São Paulo complained that their rat extermination campaign was being hampered by the popularity of Mickey Mouse who had led children to regard rodents as their friends. In spite of control campaigns, rats will always flourish so long as man persists in littering his environment with discarded food and edible garbage. In England and Wales, in 1973, it was estimated that rats infested about 1 per cent of urban properties, 50 per cent of farms and 60 per cent of sewer systems. The influx of foxes and kestrels into British cities over the last few years, however desirable from the naturalist's point of view, is a telling commentary on the habits of the citizens. Today, rats are less numerous around farms than they were a few years ago, but a visit to a picnic spot or roadside litter bin will show that country rats too are enjoying the fruits of man's affluence.

8

PATTERNS FOR SURVIVAL

Since Darwin's time, nearly every *Introduction to Biology* has used rodents of one kind or another to illustrate the concepts of food chains and pyramids of numbers, invariably placing them in the least enviable position in the chain or pyramid. With so many birds of prey, carnivores and snakes depending upon rodents for food, there is a tendency to regard such universal prey animals simply as animated food parcels, unable to defend themselves and, by prodigious feats of multiplication, continually striving to satisfy the appetites of numerous predators. This concept overlooks the fact that often the rodent link in the food chain is made up of not one, but perhaps a dozen, species and the mortality suffered by each member through predation may be so modest that reproductive rates no greater than our own are sufficient to perpetuate the species. As for rodents being defenceless, anyone who handles them will soon learn that incisors can be put to other uses than feeding. Even among species which bear the brunt of predatory attack in certain areas, natural selection has equipped each individual with some mechanism for survival.

The first lines of defence practised by most rodents are their unobtrusive habits whenever they leave the burrow or shelter. There are exceptions. Tree squirrels have little need for concealment and unless they have been extensively hunted, usually show little fear of man or beast. Many aquatic species are similar; the European water vole is short-sighted as well as hard of hearing, and it is easy to approach within a metre of an animal busily feeding; the normally alert brown rat, by the waterside, will also tolerate a close observer, relying on its swimming ability to make a rapid escape. The techniques of concealment are well demonstrated by rats, mice and voles. In most parts of the world they are the most abundant wild mammals, yet they are seldom seen, except when dashing across a road.

With the exception of some tree squirrels, the rodent's coat is generally a nondescript grey or brown, shades which give some protection, whatever the surroundings, and do not restrict the animal to a single habitat. There are a few recognisable trends towards more specialised forms of concealing colouration. In addition to the light fawn or grey coats of desert species, cryptic colouration is well demonstrated by the bold spots and stripes of several species of squirrel-like rodents. Woodland species are frequently striped, while spotted markings are more prevalent in ground squirrels which inhabit open areas. Such animals are quite conspicuous when moving, but when alarmed, they 'freeze', pressing their bodies to the ground or close to a tree trunk, and apparently melting into the background. The suslik is indistinguishable from stony soil, the flying squirrel (*Pteromys*) can be mistaken for a piece of bark, and the African ground squirrels and American chipmunks blend with fallen branches. The black-and-white stripes of the chipmunk even extend down the sides of the face to mask the eyes. There are a few records of both African and American squirrels actually feigning death. Professor Ewer has described how a squirrel which was brought to her in West Africa, fell over apparently dead when she put her hand in its box; it remained for two minutes with its eyes shut, mouth open and no visible signs of breathing. The gundi, a hamster-like rodent of north Africa, employs a similar technique; one captive has been known to remain in a state of paralysis for twelve hours.

Several genera of rat-like rodents have striped markings, but with one exception, they are confined to the Old World. Three genera of striped mice occur in the African grasslands (page 107). *Rhabdomys*

Fig 21 *Chipmunk*

pumilio has alternating black and yellow stripes on the back; several species of *Lemniscomys* have series of bright yellow stripes or spots over the whole of the upper surface, and *Hybomys trivirgatus* has three black stripes. All these forms are diurnal, and it is probable that by breaking up the outline of the body, the markings aid concealment from birds of prey. *Rhabdomys* is particularly successful, being common on most of the high plateaux of southern Africa; although it is characteristic of grassland, most of the African plateaux were forested as recently as 500 years ago. The bright spots of some *Lemniscomys* seem more suited to the dappled light effects of woodland, and several species are most common in forest clearings. It is also significant that *Hybomys* only occurs in the forested belt extending from West Africa, across the Congo to Uganda. In South America, two genera of agouti-like rodents have light-spotted markings; the paca (Figs 22, 23) and the pacarana. Both are inhabitants of forests.

Rat-like rodents with a single stripe have a far wider distribution than those with multiple stripes. Many of them are climbers, including the birch mice and striped field mouse of Europe and Asia, and the pigmy climbing mice of Africa. A few striped forms are not associated with woodland. The steppe lemming inhabits steppes and semi-deserts yet has a distinct black stripe; so has the striped hamster, *Phodopus songorus*, which is becoming a popular pet (page 108). The markings of these non-woodland species may have had some significance in the past, before the climatic and vegetational changes associated with the ice ages. Anyone who keeps a striped hamster must notice how readily it climbs the wires of its cage, twisting and curling its tail as if it was a long prehensile appendage instead of a pathetic remnant; perhaps it is recapitulating the actions of tree-dwelling ancestors. In the New World, the only rat-like rodent with a dark stripe down the back is the Mackenzie varying lemming of Canada, but even in this species, the stripe is normally present only in the young. If one accepts that stripes are adaptive to wooded habitats, the confinement of striped forms to the Old World supports the view that at least the majority of rat-like rodents originated there, and only after the disappearance of northern forests did some, like the lemmings, cross the polar land bridge and pour down into North America. The markings of lemmings, whether a single dark stripe or the bold black-and-brown pattern of the Norwegian lemming, are of little consequence when the animals live under the snow. The adult varying lemming is cryptically coloured at all seasons; in summer it is grizzled brown or grey but the winter pelt is entirely white. It has been suggested that such colour change, unique among rodents, is advantageous because this species feeds in higher and more exposed regions than any other. In such regions there may be slight snowfall and the animals must forage on the surface. Apart from its colour, the flattish body of the lemming aids concealment by casting no shadow.

While concealment is a useful protection against aerial attack,

Fig 22 *Paca, an inhabitant of the tropical forests of South America*

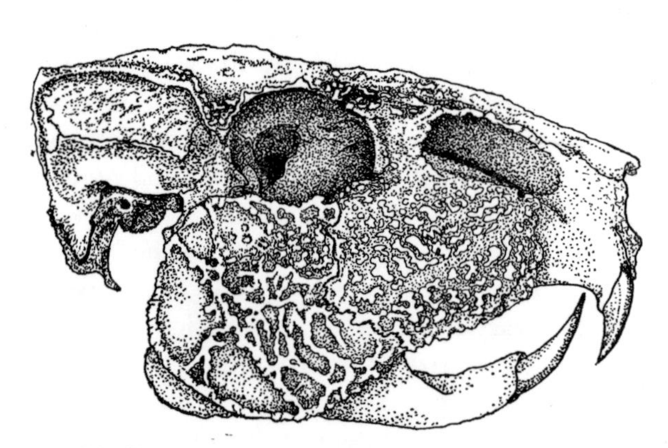

Fig 23 *Skull of paca. The cheek-bone is greatly inflated and the whole surface is ridged, giving the skull a reptilian appearance*

against predators with a keen sense of smell it is useless. When concealment fails, the next survival method to be tried is flight. In its home range, the terrestrial rodent is reasonably secure, being so familiar with its network of pathways that no time is lost in deciding which way to turn. Desmond Morris has described how the green acouchi of South America deliberately runs back and forth along definite routes in its territory to beat pathways through the undergrowth, any loose stones and sticks being picked up in the mouth and removed. The fleeing acouchi behaves like a hare, changing course frequently, and finally 'freezing' in a sheltered spot. Rodents living in open desert areas are especially dependent on speed. The great jerboa of the USSR can reach 48 km/h, and the Australian desert mouse (*Notomys*), besides being a fast runner, can also jump sideways.

When captured by a predator, the rodent's chances of survival are slight, but in some species they may be considerably increased if it is caught only by the tail. Many small boys have picked up a pet by its tail, only to find themselves ruefully contemplating an empty sheath of skin between their fingers. Any animal, when suspended in such an undignified manner, naturally twists about violently in an effort to get free. If the tail-skin is loosely attached, it may well succeed. In several species of rats and mice there is a reduction in the number of connective tissue fibres in the tail, also weak zones in the dermis where the skin easily separates. A rodent which loses its tail-skin appears to suffer little discomfort, bleeding is slight and the injured portion eventually withers and fall off. Tail-slipping is a well-known characteristic of the European field mouse (*Apodemus*) and the Florida mouse, *Peromyscus floridianus*. The American mammalogist, Layne, kept some Florida mice with three different species of snakes which are probably their natural predators. He found that 7 per cent of the mice which were captured by snakes were initially seized by the tail; also that about the same percentage of wild Florida mice had part of the tail missing, so had presumably narrowly escaped at some time during their lives. The deprivation of the tail would naturally be a severe handicap to any arboreal rodent, and the tails of most climbing species are securely attached. There is however, at least one species of African dormouse (*Graphiurus*) whose tail has not just one line of weakness but a whole series at approximately 8mm intervals along its entire length, the segmental arrangement resembling the stems of certain plants.

With the exception of man, predatory mammals kill only for food, and probably learn to ignore unpalatable prey. So far as we know, all rodent flesh is quite palatable, but some species, through special modifications of the fur, make an unattractive meal for any animal which bolts its food whole. In most rodents the fur is made up of soft insulating hairs which are interspersed with long, tough guard hairs which help in waterproofing and protecting the skin from normal wear and tear. The toughness and form of the guard hairs vary according to species and to the part of the body they protect. As one would expect, the types of hair, and their density, reflect the animal's habitat and way of life. A few snippets of fur from any rodent can provide a classroom with hours of useful study. Grooved types of guard hair occur in many species, including guinea pigs, lemmings, field voles and grey squirrels; the function of the grooving is conjectural, but it probably increases insulation and makes a strong and flexible structure. In the African grass rats (*Lemniscomys* and *Rhabdomys*), also in the harsh-furred rat (*Lophuromys*), the grooved hairs are almost bristle-like. In the spiny mouse (*Acomys*), evolution has proceeded a stage further and the coat consists mainly of grooved spines (page 108). These spines are soft enough for anyone to handle the animal with impunity, but there is no doubt that their passage down a digestive tract would be as painful for the swallower as the swallowed.

Owls are the most important predators of nocturnal rodents, and it is easy to study their dietary preferences by examining the indigestible parts of their food such as bones and fur which are regurgitated in the form of pellets. Among the 1,353 rodent skulls which I extracted from barn owl pellets in Malawi, only one belonged to a spiny mouse; since the species was common in the vicinity of owl roost, the spines must act as a deterrent to these predators. The

Page 125 (*above*) Crested porcupine, *Hystrix*; (*below*) American porcupine, *Erithizon*

Page 126 (*above*) Pouched rat, *Saccostomus*, carrying young; (*below*) acacia rat, *Thallomys*, with young attached to teats

possession of spines is associated with an extremely delicate skin, and the tail-skins of spiny mice slip very easily. Of the 162 African spiny mice which I collected, 18 per cent of the males and 30 per cent of the females had shortened tails. In Panama, Fleming collected 637 spiny rats, *Proechimys semispinosus*, and noticed that 18 per cent had incomplete tails; in this species the breakage always occurred at the centrum of the fifth vertebra. It seems unlikely that in such well-protected species as spiny rats and mice, this high percentage of tail amputations is caused by owls; many are probably caused by fighting among the rats themselves, for it is rare to see a captive spiny mouse with an intact tail. The African cane rat, which is a member of the porcupine-like sub-order of rodents, also has a coat of soft spines, but it is doubtful whether they afford much protection. This large species is well known in many parts of southern Africa through its habit of noisily chewing sugar cane and other grasses during the day. Because of its size and succulent flesh, the cane rat is eagerly hunted by dogs and men; in some areas it also forms an important part of the leopard's diet. Like the spiny mouse, it has a very delicate skin, freshly trapped specimens often tearing themselves to pieces through dashing against the wires of the pen or cage.

Defensive spines reach the peak of development in the porcupines, whose quills are sufficiently strong and sharp to pierce the skin of any predator. There are two families, one confined to the Old World and one to the New. Of the four genera of Old World porcupines, those popular inhabitants of zoos, the crested porcupines (*Hystrix*) (page 125), are the largest, and attain a weight of 27kg. The popular name is derived from the long wiry hairs on the neck which can be erected to form a crest. Members of this genus occur in Africa, Europe and south-east Asia. Like the rattlesnake, the crested porcupine is provided with an audible warning device on the tail. Its 'rattle' is formed from what are probably the most specialised hair structures in the animal kingdom. In the young animal, the tips of certain thin-walled tail quills break off soon after emerging from the skin, leaving open wine-glass structures attached to the skin by slender stalks (Fig 24). When irritated, *Hystrix* vibrates its tail and

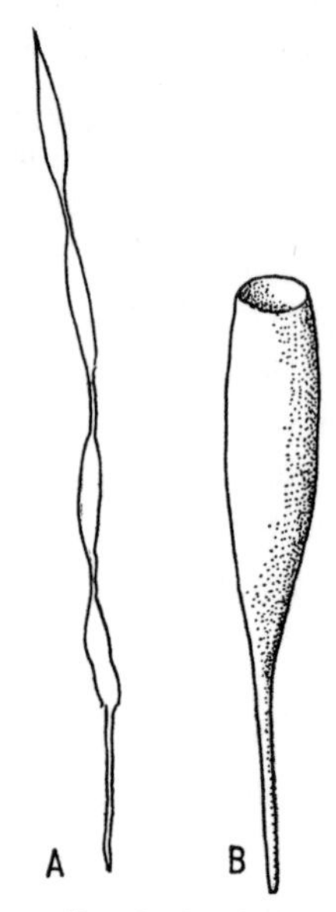

Fig 24 *Rattle quills of* (A) Atherus; (B) Hystrix

the rattle quills shake against others, to sound like the maracas of a Latin American band. If the enemy is still undeterred, the porcupine runs backwards in an attempt to impale it with the pointed quills of the rump. These are the quills which are often used as fishermen's floats; they are round in section and consist of a pith of large air cells surrounded by hard, horny layers. Smaller spines on the sides of the body are grooved like those of the spiny mouse, possibly acting like the rain-water guttering which they so closely resemble. In the smaller brush-tailed porcupines (*Atherus*), forest species of Africa and Asia, the rattle quills differ from those of *Hystrix*; each one consists of a series of up to six enlarged air cells connected by thin strands, the whole tuft making a most efficient rattle. The protective spines are also quite different, being short and having a groove both above and below, while the lower edges are furnished with small scales. If they serve no other purpose, these scales must carry infection into any wound which the spines inflict (Fig 25).

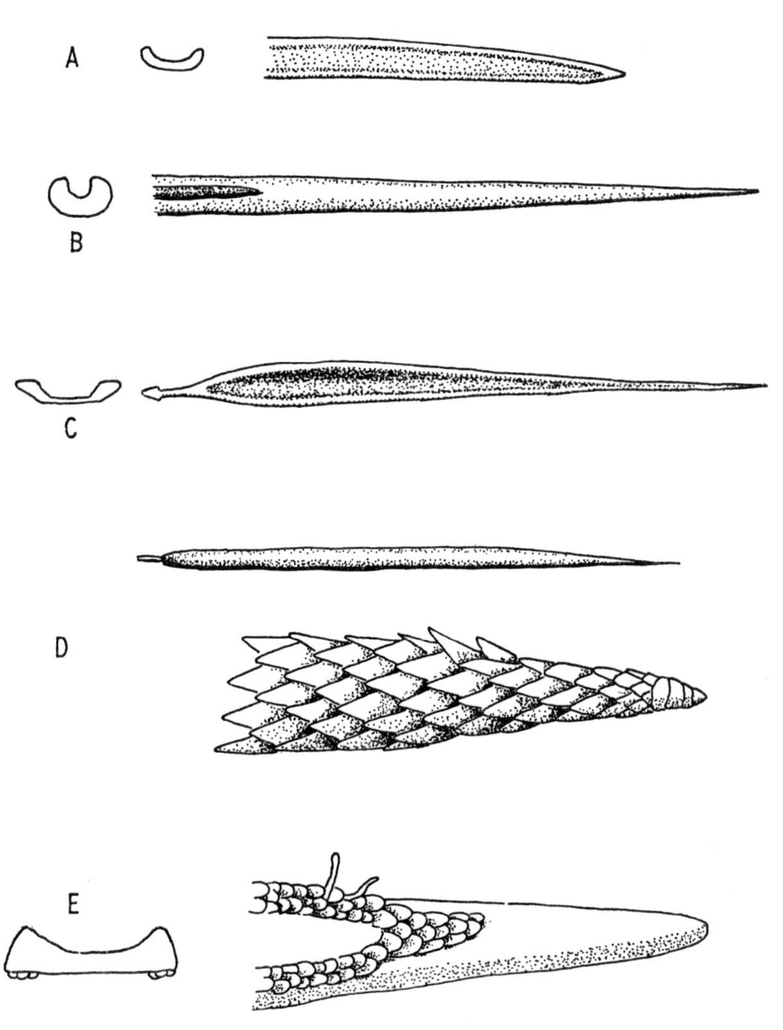

Fig 25 *Protective spines:* (A) *spiny mouse*, Acomys (*tip only*); (B) *cane rat*, Thryonomys (*tip only*); (C) *spiny rat*, Echimys; (D) *porcupine*, Erethizon (*whole quill and tip greatly magnified*); (E) *porcupine*, Atherus (*tip of quill, lower surface*)

The quills of the North American porcupine, *Erethizon dorsatum* (page 125), are only about 30mm long but are more sophisticated weapons than those of other genera, for the tips are covered with hundreds of minute barbs. *Erethizon* cannot shoot its quills, but when attacked, it may lash out with its tail and leave a few dozen sticking in its adversary. The victim's muscle fibres naturally twitch at being used as a pin-cushion, and engage the barbs, slowly but surely drawing the quills deeper into the tissues. Quills penetrate at the rate of about 25mm a day. Porcupines can be a nuisance to ranchers since cattle delight in chasing the rodents, sometimes receiving so many quills in their faces that they are prevented from feeding unless dequilled. With such an armoury, it is not surprising that the porcupine is a relatively long-lived rodent; one individual tagged in Michigan was recovered when at least ten years old. Analysis of 3,000 wolf stomachs showed no trace of porcupines being eaten, but a similar survey of the diets of other carnivores in British Columbia, showed that a quarter of the fishers and foxes examined had eaten porcupine at some time or another. For some unknown reason, porcupines are the preferred prey of the fisher. This small carnivore's method of attack is to circle round its prey, dodging the tail and delivering rapid bites to the head. When the porcupine is down, the fisher turns its attention to the soft, unprotected underside. Eventually nothing is left of the rodent, apart from a handful of quills. In some parts of the USA, the great increase in numbers of porcupines over recent years has given concern to foresters, owing to the damage caused to timber. In some forests, porcupine densities of twenty-five per sq km have been recorded. Between 1961 and 1963, sixty-one fishers were taken from Minnesota and released in Michigan, in an attempt to reduce porcupine numbers. Although a coat of quills provides immunity from most other predators, it is not entirely without disadvantages. Should a porcupine become too ardent in its courtship, it is liable to be pierced by its partner's quills![140]

Spines have evolved in two genera of squirrel-like rodents, the spiny pocket mice of Mexico and Central America, but one might imagine that tree squirrels, palatable as they are, could live securely in their wooded habitats without any special protective devices. It has been suggested that several species of Malaysian tree squirrels mimic the distasteful tree shrews with which they share their habitat. These squirrels are remarkably similar to tree shrews in shape, size and colour; the only obvious distinguishing feature is the more pointed head of the shrew. Both tree shrews and squirrels are diurnal, and are so alike in the wild that the Malaysians apply the name 'tupai' indiscriminately.

Small size is often an important factor in survival, for it allows an animal to elude many predators by diving into the nearest hole or crevice when danger threatens. In many parts of the world, such holes often accommodate disconcerting bedfellows like scorpions, which are as lethal as any predator. I often encountered scorpions while digging for rats in Africa; in one low-lying area there was one in every few square metres of soil. In some countries they cause more human fatalities than snake-bite. In the Mexican Republic, during periods 1940–9 and 1957–8, 20,352 people were killed by scorpion stings. Several species of rodent live in scorpion-infested areas, but few studies have been made on their means of co-existence. The round-tailed ground squirrel, *Spermophilus tereticaudatus*, which occurs in Arizona and Mexico, has been found to have considerable immunity to scorpion venom; weight for weight, it is over thirteen times more resistant than the laboratory mouse.[163] Several species of small rodent, including the American harvest mouse, are sometimes eaten by scorpions, but the grasshopper mouse, which has roughly the same distribution as the round-tailed squirrel, manages to overcome and eat scorpions longer than itself. It is not known whether the grasshopper mouse is immune to scorpion venom, but B. Elizabeth Horner and her colleagues studied the species in Nevada, and from scat analysis, found that scorpions had been eaten by eighteen of the forty-nine mice studied. When a 74mm scorpion was placed in a mouse enclosure, it was immediately pursued and the mid-tail segments were attacked repeatedly; once the sting was immobilised, the scorpion was decapitated and the flesh was eaten. Grasshopper mice can even overcome whip scorpions which spray a concentrated solution of acid. The method of dealing with such dangerous prey is an interesting example of natural selection, for the mice could hardly learn to avoid the sting through experience.

The nocturnal feeding habits of many species of rodent may be an adaptation to climate or to avoid diurnal predators, but little is known about the effect of moonlight on surface activity. Grasshopper mice kept in an open pen were found to be greatly influenced by the phases of the moon. They were most active between the last-quarter and first-quarter moon, and least active at full moon. During the first quarter, peak activity was in the evening, and during the last quarter it was in the morning. In this species, the activity pattern may be associated with avoidance of predation by owls, or with rhythmic activities of its arthropod prey.[77]

The sight of a pet hamster or mouse devouring its new-born young does not give a good impression of the rodent's parental qualities, yet in those forms whose young are born in a helpless state, parental care makes an essential contribution towards survival. Generally, rodents are most solicitous parents, carrying the whole litter to a place of safety if the nest is disturbed. The infanticidal tendencies of some captives are usually due to premature births induced by careless handling, or to some other condition of their captivity. Directly a baby is born, it is licked thoroughly by the mother; if this produces no reaction, it is eaten along with the placental membranes. Most mammals consume the birth membranes, so keeping the nest clean and obtaining nourishment at the same time; in the case of the beaver, the male joins in the rather messy feast. The maternal eye is an uncritical one, and rodents usually adopt any strangers which are introduced into the litter. Entire litters of striped hamsters have been exchanged without exciting the parent's notice. Another parent hamster was quite unconcerned when her litter was suddenly doubled from three to six; she just picked up the newcomers and deposited them in the nest with her own babies. Even the most maternalistic human would be expected to raise an eyebrow if a mere 1kg stripling apparently resulted from her labours, yet one of my spiny mice readily adopted an infant pigmy mouse just one-third the size of each of her own two babies. Some of the problems of tree-dwellers have already been discussed, but not least is their difficulty in quickly evacuating the young from the nest. The African Acacia rat (*Thallomys*) usually has two young which remain permanently attached to the mother's teats for about the first fortnight of life (page 126). They do not restrict the mother's movements since the teats are at the base of the abdomen; the babies themselves seem to suffer no discomfort as they swing between her hind legs or drag over the branches during nocturnal forays several metres above the ground. Similar behaviour has been reported in the banana rat (*Melomys*) of New Guinea.

As we know from the population dilemma facing mankind today, any successful species faces the danger of over-exploiting the resources of its environment. Having been taught at school about biblical plagues, and possibly somewhat awed by the information that a pair of rats is theoretically capable of multiplying to 10,000 in a couple of years, it may seem rather pointless to pursue the subject of population control among animals which 'Boil over in fields and villages', or hurtle over cliffs, apparently to commit mass suicide. Contrary to the general belief, the reproductive proclivities of most species of rodents are quite restrained and often only modest numbers of offspring are produced. In fact, the African spring hare may be taken as a model of reproductive restraint since it only produces a single young each year; the South American paca is a close rival but usually produces two young per year. Of the twenty species of rats and mice found in Malawi, only five produce litters with more than four young. To estimate the reproductive potential of an animal, it is of course, necessary to have some idea of the number of litters produced by a female during her lifetime. In many species, this information can be obtained through dissecting the female's body. Whenever an embryo develops in the uterus, an associated pad of placental tissue is formed and this pad, or placental scar, may remain throughout the rodent's life and her past childbirth history can be revealed by a few snips with a pair of scissors (Fig 26). The counting of the scars is not an accurate way of determining the number of births since some embryos may be resorbed, or, if large numbers of scars are present, some may fuse together; even so, in many species, the scar counts can give a reasonable idea of the total number of past implantations. As their appearance varies with age, they also indicate the number of litters produced.

The method cannot be applied to all species; two species of American voles have been shown to retain their scars for less than fifty days. In the African rats examined, the placental scars apparently remained throughout life and indicated that only six species regularly produced more than two litters in a lifetime. Few females

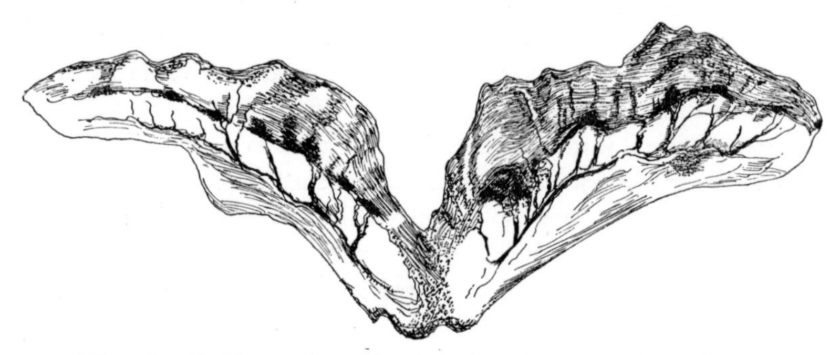

Fig 26 *Oviducts of rat showing about fourteen placental scars*

produced more than ten offspring during their lives, most of them about six—roughly the same as the average human family before birth control became widely practised. There was one exception. This was the multimammate rat, *Mastomys natalensis*, a significant common name since mammae are only ornamental in one mammalian species, and the multimammate rat has up to twenty-two. The litter size of this species ranges from seven to seventeen, and several litters may be produced in a lifetime. With such a reproductive potential, it is not surprising that the species sometimes reaches plague proportions in many parts of Africa, destroying cotton, ground nuts and other crops. A few other African rodents such as gerbils and grass mice also show marked fluctuations in numbers, but few studies of them have been carried out.

In the northern hemisphere, rodent populations have been studied for many years and it has long been known that in certain regions the numbers of some species gradually build up to a peak then suddenly decline. Sometimes the population cycles occur so regularly that peak years can be predicted with reasonable certainty. In agricultural districts, a peak year for rodents may be an economic catastrophe, but in others it may be quite the reverse. That pioneer of ecology, Charles Elton, through checking old trading returns and records of the Hudson's Bay Co and Canadian mission stations dating back to about 1834, proved beyond doubt that the abundance of foxes was closely linked with the numbers of voles and lemmings; the populations of both carnivores and rodents showing a cyclic variation every three or four years. In the years following a crash in rodent numbers, the predators either die or disperse. And many arctic birds of prey fly south to be gleefully recorded by birdwatchers, or in less enlightened times, to be shot by collectors. Elton records that in the winter of 1926–7, North American taxidermists were flooded with orders to mount snowy owls. One firm in Boston received 143 owls and it was estimated that at least 5,000 birds were killed during their exodus from the northlands; many of the snowy owls seen in museums date back to this massacre. It will never be known how many owls dodged the guns or remained in the north, but rodent numbers must have been prodigious to support such a host of predators since 5,000 snowy owls would eat about 14,000,000 voles in a year.

In the circumpolar tundra, the snowy owl feeds chiefly on lemmings, those vole-like rodents so often visualised as following blindly behind their leaders, like motorists hurtling along a fog-bound motorway, and bent only on self-destruction. During a lemming year there are indeed mass movements, when large numbers sometimes converge on coasts, and finding themselves on cliff-tops, may dive without hesitation into the sea and swim towards the horizon. During an irruption in numbers, the lemming completely changes its habits. Instead of living quietly alone, it is seized by a kind of wanderlust, living a nomadic life on the surface and moving down the mountain slopes. There is no organised emigration, no leaders, and no definite route but usually a general compass direction is followed which tends to remain the same for a particular area. In the valleys, the trickle of individuals merge to form a stream. Eventually there is a flood of animals all moving in one direction, cropping the mosses and grasses, and devouring the bodies of fallen comrades. While on the move, lemmings are far from timid, standing their ground when threatened by man or dog, and even attempting to bite. De Kock and Robinson observed an irruption in Sweden in 1963, and described how lemmings were visible for as far as the eye could see; at one stage, forty-four were passing every minute, and on one part of the shore the animals covered nearly 300m of shoreline to a depth of 50m, the individuals being so tightly packed that it was difficult to avoid stepping on them. Lemmings are not great swimmers but can paddle along at a speed of 1km/h, and can cross any lakes and rivers which lie in their path. However, they do not plunge recklessly into raging torrents or enter water with waves more than 30mm high; those which enter the sea are almost certainly under the misapprehension that they are just crossing another lake. Far from being suicidal, the lemmings, like ourselves, are simply seeking a better land. Not all lemming movements are sufficiently dramatic to be termed emigrations. Many involve local wanderings of a few kilometres; usually, it is only when populations in different areas build up at the same time that movements are at all noticeable. Large-scale movements are not limited to lemmings, sometimes they are reported for other species, particularly voles. A Russian author, in 1908 described an irruption of social voles in the Transcaucasus, and recorded that,

> Not one spot uncovered by the tiny rodents could be seen. In some stretches we could see dark black spots, consisting of hundreds of thousands of half-dead animals collected in heaps to constitute, as it

> were, a banquet pile for their voracious brothers to polish off... The area traversed by the voles became completely black without the slightest blade of grass or clump of herbage surviving. Nothing but piles of skeletons and carcases remained as mute witnesses of the migratory wave of these small pilgrims.

Among rodents possessing a high reproductive potential, populations sometimes show spectacular increases after experiencing a few seasons of plenty, but there is some evidence that the quality, as well as quantity of vegetable food affects reproduction, at any rate in some species. For many years it has been known that plants contain hormone-like substances which can induce oestrus in mammals, and that these substances are most prevalent in young growing plants. Two American zoologists, Negus and Pinter, have thrown a new light on the properties of that homely vegetable, spinach. Spinach extract was fed to young female voles and when each animal had consumed the equivalent of 15g fresh spinach, it was discovered that the uteri were much heavier than those of controls, and the ovaries contained twice as many developing follicles. The addition of sprouted wheat to the diet had an even more dramatic effect; nine out of ten voles came into oestrus within twenty-four hours of being fed with 5g, and continued feeding led to an increase in the numbers of litters and offspring, also to a reduction in infant mortality. Day-length plays an important part in determining the breeding seasons of some species living in temperate zones; for example, the European vole, *Microtus agrestis*, breeds only during months which have over 100 hours of sunlight. In the tropics, however, most rodents still have well-defined breeding seasons, peak reproductive periods corresponding with the seasonal growth of new vegetation, either in the spring or the rainy season. Some species, such as the Merriam kangaroo rat of California, show two peaks of pregnancy, one in spring and the other in autumn; possibly the autumn peak is due to the high proportion of fresh seeds in the diet.

The South American rodent plagues known as 'ratadas' illustrate the influence of food supply on reproduction. In Chile they occur every seventeen to twenty years, and in parts of Brazil every thirty years. The 'ratadas' are usually associated with the fruiting of certain species of bamboo, and Brazilian farmers believed that every cane housed a grub which developed into a rat when the plant fruited and died. The real explanation is that the canes in an area do not shed seed simultaneously, but over a five-year period once in every thirty years. During the seeding period, enormous quantities of the rice-like seeds are produced, often covering the ground to a depth of 15cm. With such an abundance of food, the rat population builds up during the years of plenty, then, when the last plants die, the animals are forced to emigrate to cultivated areas. In Brazil, the species of bamboo concerned is the rough bamboo, *Merostachys fistulosa*; its dying canes are often infested with insect larvae, and no doubt it was these which gave rise to the quaint belief as to the 'ratadas' origin.

Many suggestions have been made concerning the causes of the crash which inevitably follows a rodent irruption. Some studies, notably those of Barnett, have shown that overcrowding of rats and mice leads to frequent conflict between individuals, resulting in an enlargement of the adrenal glands. And it has been postulated that stress leads to shortened lives and a lowering of fertility. Other workers have suggested that populations crash through shortage of food, and to high mortality caused by the influx of predators. Recently, Krebs and his co-workers carried out experiments with two species of American vole, keeping some in enclosures to compare with nearby wild populations. There was no evidence of any stress, even though the enclosed voles reached such abnormally high densities that eventually they died from starvation. Krebs's team also tried to discover why members of increasing populations grow more rapidly and reach a larger size than members of declining populations. By comparing the serum proteins of the voles, and using them as a kind of fingerprint identification system, it was found that individuals in dispersing populations were genetically different from those in resident populations. The dispersing voles were larger and demonstrably more aggressive than residents. These observations suggest that under conditions of overcrowding, a genetic change affects at least some of the vole's offspring; these individuals, with behavioural traits quite different from those of their parents, will be the ones to leave home and seek new territories. They also explain how, during irruption years of voles and lemmings, some individuals remain behind to restock the homelands.

Whatever the causes of rodent plagues, emigrations and sudden declines, such violent population upheavals are limited to a relatively small number of species, usually those which feed principally on grass. Grass is one of the few food materials which is in almost inexhaustible supply; even if overcropped, it reappears the following year. Without periodic population crises, grassland species would lack any incentive for extensive dispersal over such a homogenous type of habitat.

9

SOCIAL LIFE AND RODENT COMMUNITIES

Anyone who has kept pet rodents can appreciate the world of difference between the behaviour of a guinea pig and a golden hamster to members of their own kind. While the former settles down amicably with any number of strangers, the hamster immediately attacks the first intruder. It is no quirk of nature that has made the hamster solitary and the guinea pig gregarious. Environmental conditions influence the social characteristics of a rodent, or any other animal for that matter, in just the same way as they affect anatomical features. In turn, the extent of an animal's social contacts has a bearing on the evolution of methods of communication and other aspects of its life. Although one might not expect to find any resemblance between the behaviour patterns of solitary species and those which live in large communities, detailed studies have revealed some striking similarities.

Some rodents are more solitary than hamsters, living in separate burrows and making contact with the opposite sex just once each year. Examples are the mole rat (*Spalax*) and the African pouched rat (*Beamys*), whose isolated burrows are widely spaced in the evergreen forests. As in most animals, the female is normally the most aggressive sex, and any male is liable to suffer if his advances are inopportune. In the wild, the liberation of odour or other signals by the receptive female probably minimise this danger. As hamster breeders are aware, a male is likely to be severely injured if it is confined with a female for longer than its duties require. While mating can be limited to a brief interlude, there must be a longer period of tolerance between the female and her litter. Litter mates also must associate with one another at the commencement of their lives. In some species, such periods of juvenile contact are of short duration, and the aggressive tendencies of the young lead to rapid dispersal directly they are old enough to fend for themselves. When dispersal is prevented by captivity, some species can adapt by setting up a kind of social structure. Striped hamsters can be induced to live together fairly peaceably, and two or three generations may share the same nest. Other species can only set up a community if the cage is provided with a number of nest boxes allowing each individual to establish and defend its own small territory.

Living together involves a number of behaviour patterns and social signals. Scent, which will be mentioned in more detail later, is an important means of communication in all types of association, while touch, and sometimes sound, are particularly important in the relationship between mother and young. Excepting in those species which are well-developed at birth, the heat-regulating system of the newborn rodent may take some days to function properly. In the Mongolian gerbil, it does not become fully operative until the young are twelve days old. Being in effect 'cold-blooded', the youngster relies on its parent for warmth, and cannot survive for long if it strays outside the nest. The young of many species call loudly when they are cold, their cries inducing the mother to make a search and recover any lost wanderer. Species which are virtually non-vocal throughout adult life, may be quite vociferous for a few days just after birth. Baby striped hamsters make creaking calls, rather like a nail being pushed along the teeth of a comb; these calls often are the first intimation to their owner than a new litter has been produced, for they cease after a couple of days. Calls of this nature have been described for other species. Although the human ear cannot distinguish between them, probably each species has its characteristic call. When the squeaks of three-day-old young of two sub-species of deermice were analysed, their sonograms, or voice prints, were found to be quite different. In 1954, it was reported for the first time that laboratory rats emitted high-frequency sounds which were inaudible to the human ear. Until then, bats were believed to be the only terrestrial mammals to produce ultrasounds. In succeeding years, their production has been detected in the newborn young of laboratory mice, collared lemmings, European field mice, *Apodemus sylvaticus* and *A. flavicollis*, and the European vole, *Microtus arvalis*. Doubtless the list will grow as more species are tested. By playing back recordings of the calls made by baby field mice, Gillian Sewell induced nursing females to leave their nests and quickly orientate towards the loudspeaker, although none of their young were missing.

There are few records of male rodents playing any part in the care of their progeny; probably the majority never see them since they are driven away by the female soon after mating. Even when females are not hostile, many breeders prefer to remove the male because some forms, such as guinea pigs and voles, come into oestrus within an hour or so after giving birth, and remating at this time may impair the subsequent health of a mother required for showing purposes, and also disturb the newborn young. The males of gerbils and deermice, at least in captivity, are allowed by their mates to share the breeding nest and help in keeping the litter warm; they also retrieve any young which stray. In captive striped hamsters which have become socialised through confinement, the presence of an infant arouses the interest of all members of the community. If it is picked up by a juvenile, it passes from mouth to mouth until rescued by an adult. One experimenter tested the maternal response of 115 white mice of both sexes by placing a one- or two-day-old pup outside the nest; all the mice retrieved the pup and only two failed to lick it.[112]

A few species, including the chinchilla, beaver, springhare and grasshopper mouse seem to maintain the pair bond throughout the breeding season, possibly throughout life. The southern grasshopper mouse has an elaborate courtship ritual, the pair chasing each

Page 143 (*above*) Prairie dog; (*below*) viscacha

Page 144 A very large beaver

other, standing on their hind legs, rubbing noses, and chirping the whole time. After mating, both partners eat and sleep together with complete equality, but during the female's confinement, the male lives outside the nest for a while. Once the litter has been born, he becomes subservient, delaying his mealtimes until she has fed—a form of etiquette which takes into account the importance of a well-nourished mother. The male often cares for the litter while the female is away feeding, also retrieving straying youngsters, covering them with nesting material, and even washing them.[70] Studies of the northern grasshopper mouse revealed that both partners helped in making the nest burrow, the male digging out the sand and kicking it to the female behind who cleared the spoil from the burrow. Juveniles were also seen trying to help their mother in kicking out sand.[134] Young beavers remain with their parents until they are about two years old; they are particularly useful in assisting in the building of dams and lodges. Family parties of beaver never contain more than fourteen individuals; too many get in each other's way and become aggressive.

The family group offers many advantages over a solitary existence; not only are the tasks of constructing shelters and burrows eased, but the young, if orphaned, are protected by other members of the family. A member of a community also has no problem in finding a mate. There are, however, two obvious reasons why so few species have been able to enjoy the benefits of communal life. Even two animals living together need twice as much food as one, and if each has to travel further to find it, they are more at risk than the animal living alone. A predator is also more likely to locate a group than an individual and to keep returning until the population is decimated.

Even where the environment offers ideal conditions and can support a large community, there is no evidence that any gregarious species of rodent can live with its neighbours in a kind of co-operative commune. An egalitarian society still remains a pious hope for man, but for rodents it is a non-starter. Several detailed studies have shown that a rodent community survives as much by the anti-social behaviour of its members as by their tolerance to each other. These anti-social tendencies, and systems of social rank, prevent population pressures causing a community to degenerate into a slum. Food and shelter are not the sole considerations in determining the spacing of a community. There must be room for each female to rear her young without disturbance from neighbours. Dr Swanson of Birmingham kept two pairs of golden hamsters in an enclosure of about 6sq m to find the rate of population growth. Although the species is notoriously anti-social, it was surprising to find that after eight months there were still only four hamsters, despite the fact that nine litters had been produced. The low survival rate was believed to be due mainly to too many animals sharing the nest.

From the few studies carried out, mainly on domestic species, it seems that the gregariousness of some rodents is comparable to that in primitive human societies; each group or tribe consisting of a number of small family units with a greater degree of tolerance between their members than to other families. If house mice are released into an enclosure, the first few hours are spent in exploring their new surroundings, there is rarely any fighting. For the next two days, fighting is intense, especially among the males as each tries to establish its territory. Thereafter the combats subside. One series of observations showed that the colony remained peaceful until the oldest litters were 120 days old and tried to establish their own territories. The numbers of mice in each territory varied considerably, one housing 7 individuals, another, 168. Each territory was occupied by a breeding unit consisting of a dominant male, 2–5 females, up to 3 subordinate males and a number of juveniles. Although males were restricted to their own territories, the females, which also held territories of a kind, were able to use those of neighbouring males.[125] In another experiment in which colonies were allowed to become severely overcrowded (about 150 mice in 0·5sq m), it was found that up to 56 per cent of the males received bites or scars through fighting, but very few, and sometimes none of the females were injured; some pregnant females successfully defended their territories against males. A strange mouse was immediately pursued by all the residents; even pups which were barely out of the nest were able to drive away an intruder.[36]

Barnett's observations on laboratory colonies of brown rats revealed that groups of males lived together without conflict. But immediately females were introduced, the members of the celibate fraternity turned on each other and fought so viciously that often there was only one survivor. Apparently the males did not fight for actual possession of the females; it seemed as if the mere presence of the opposite sex led to increased excitement and frustration which found its outlet in extreme aggression. In colonies where the feminine influence was less traumatic, the males settled down by means of a social rank system. There was a top tier of dominant males, a middle tier and a lower tier. Middle-ranking males managed quite well in a subservient position and usually slept together, but members of the lowest group lost weight and died in a short time, sometimes within ninety minutes. As there were few signs of physical injury, these were presumed to have died from some form of stress or shock. In a wild community, any low-rankers would be able to escape, for fighting to a finish is the ultimate folly so far as the survival of any species is concerned, and it is normally prevented by elaborate behavioural mechanisms. Just as a man cringes before his superior to avoid retribution, a submissive rat crouches in front of a dominant member of the group, sometimes actually crawling underneath it. Aggressive encounters among brown rats are accompanied by the emission of ultrasounds by both animals. Long pulses are characteristic of submissive animals and short pulses of higher frequency are made by the dominant. Provided one animal emits the long 'surrender' signal, the aggressiveness of the other soon diminishes.[135] Future studies might well show that communication by ultrasound plays an important part in the social lives of other species. Although its significance has yet to be established, it probably has some survival value if inaudible to predators. It is still not known how the sounds are transmitted or detected; perhaps, as in bats, dolphins and oilbirds, ultrasonics evolved as a means of getting about in darkness, and the communication function is only secondary.

Animals which live together must be able to keep in touch with each other. But since any means of communication is liable to alert predators, communal life is only open to those which are either fairly secure from attack, or have an early warning of an enemy's approach. There are a few gregarious, nocturnal forms such as gerbils which inhabit open, rather sandy areas, where they can maintain contact by sight and sound. Owners of pet gerbils sometimes may be disturbed by a rhythmic drumming coming from their pet's cage. The sound is produced by the gerbil rapidly vibrating its hind feet against the floor. Although the drumming is obviously a social signal of some kind, it is hardly intended as an alarm because captives sometimes drum after food has been placed in the cage, and in any case a threatened animal is unlikely to wait to beat a tattoo. Several other forms use similar mechanical means to communicate over a distance. The kangaroo rat drums with one hind foot at a time, and deermice drum with the front feet. The wood rat can make a rattling sound by waving its tail in a circular motion and striking it against twigs. The rattle is audible 15m away, and has been seen to warn away other wood rats, and also to drive deermice from a wood rat's house. Mechanically produced alarm signals have been described for a few species. The South American acouchi stamps its feet before running away, and a disturbed beaver slaps its tail against the surface of the water before diving.

With the exception of the morse key and Congo drum, no mechanical device is as effective as the voice in the speedy transmission of a variety of signals. Unlike the foot-stamp, it doesn't hinder the animal's flight, and can be used anywhere, even in water. Vocal signals only are effective in dense vegetation, and one might imagine that the risks inherent in their use would prevent any sociable rodent living in such an environment. However, in South America, several species of communal rodents live in jungle and grassland, members of each community keeping in touch by calling to each other. Among its other peculiarities, the rodent fauna of this continent is notable for its noisiness. All manner of calls are represented, from the incessant chirps and squeaks of guinea pigs, bubbling calls of tucu-tucos, the mournful courtship cry of the coypu, to the varied clicks and grunts of the capybara; a shot capybara has been described by Barlow, as uttering yowls and screams reminiscent of a dog in pain. Such noisy species could only evolve in the comparative absence of predators, and it was at a recent date in evolutionary history that carnivores crossed the Central American land bridge. It is likely that the calls of these species serve several functions, but no detailed studies have been made. The Bahamian hutia, which is nocturnal, keeps in touch with its neighbours by squeaking, while the chattering notes of a captured animal serves to attract any others in the vicinity. A tape recording of the chatter produces the same effect.[29]

The majority of colonial species are diurnal, and live either in the safety of trees, or in regions of short grass and open mountain sides. In these situations there is safety in numbers, for the co-operative efforts of several pairs of eyes enable each animal to feed in security. Diurnal species are not only easier to observe in the wild, but are far easier to understand than nocturnal forms which rely mainly on other senses. The latter normally vanish down their holes at any unfamiliar occurrence, whether dangerous or not, but diurnal rodents are far more discriminating and use a variety of visual and oral signals to communicate with each other. Warning marks like the white rump markings of some antelopes, are found in a few South American species. The mara has a vivid white rump patch which serves as a danger signal to its fellows when it flees. Some species of agouti have long rump hairs of contrasting colours which may serve a similar purpose or be used for sexual display; there are white-rumped, black-rumped and orange-rumped species. As part of their social repertoire, many species of squirrels are able to flatten the bushy tail to display conspicuous banded markings; in others, the ear tufts and striking body colours may be used for communication purposes. Extraordinary as it may seem, body markings may play a part in the social lives of nocturnal species. The golden hamster has a black mark on each side of the chest, and it has been found that if an animal has its spots intensified by dyeing, it wins any fights against normal hamsters, irrespective of whether it had lost in previous encounters.[61]

In Europe, Asia, North America and Africa, the most familiar colonial rodents are ground squirrels and marmots. Since tree squirrels, with their keen vision and diurnal habits have evolved elaborate social signals, it is not surprising that many of their ground-living relatives have adopted the colonial way of life. Colonial forms are not always sociable, even to members of their own families. The Californian ground squirrel spends most of the year in communal burrow systems some 61m long with about twelve exits, but according to Jean Linsdale, every individual has its own preferred exit, and in the breeding season each female retires to her exclusive breeding burrow. The American rockchuck or yellow-bellied marmot, *Marmota flaviventris*, occurs in rocky areas of the western states, and lives in family groups of usually one male and several females. One colony which was studied, consisted of a male, eight adult and two yearling females. Sexual attachments seemed quite relaxed, there was no rivalry between the resident and any male which strayed in from a neighbouring colony, and a female sometimes mated with several males. For the first three weeks after the colony had emerged from hibernation, the male was the first to appear in the morning and the last to leave at night. Not for him the hectic courtship chase, for he stayed put and attracted the females by conspicuously wagging his tail. By the fifth week the male was understandably spending the greater part of the day in its burrow.[2] In this species there is a kind of peck-order among the females, subordinates tending to be driven away from the centre of the colony. The woodchuck, *Marmota monax*, of the eastern USA and Canada, usually occurs in open woods and fields. It is rather solitary and the male is a wanderer, visiting females for procreative purposes.

Marmots and ground squirrels are especially solicitous of their young. There is a record of a yellow-bellied marmot actually chasing a marten which approached too near its colony. A female thirteen-lined ground squirrel has been seen to place herself between the litter and an attacking snake; after the snake had been driven away, the young were led back to the burrow by a roundabout route.[99] The Uinta ground squirrel, *Citellus armatus*, of Utah and neighbouring states, tends to be unsociable although it sometimes lives in large colonies. Males fight during the mating period, and pregnant females are intolerant of all members of their kind and take little notice of their young once they appear on the surface. This species is known to use a repertoire of six different calls in their threat postures. Two of these calls are also used when a predator is seen; a chirp is made at the approach of a hawk, while a ground predator elicits a churr. Either call alerts all squirrels in the neighbourhood, but there is no evidence that they produce different responses.

In the great plains of North America and Asia, a few species are able to exploit fully the benefits of colonial life by living in very large communities; the hundreds of pairs of eyes prevent a predator approaching undetected. It has been reported that anyone entering a colony of Siberian marmots is immediately surrounded by a ring of sentinels, which stand on their hind legs and whistle continuously. As the man moves, the ring moves with him. Although alert, the marmot is extremely inquisitive. This trait is sometimes exploited by hunters who, by turning their clothes inside-out, rolling on the ground and waving their legs, manage to get within gun-shot range of their bemused audience.

The prairie dogs, so-called because of the dog-like bark, are close relatives of the marmots but are limited to the central USA and northern Mexico. Two species are usually recognised; the rather solitary white-tailed species, and the colonial black-tailed. At one time, large dog-towns were common throughout the prairies, but nowadays very few exist outside the national parks. Most of our knowledge of dog-town life is due to J. A. King who spent three summers and a winter in the Black Hills of Dakota, studying a dog-town which covered about 30·4 hectares and contained nearly 1,000 inhabitants. A town is divided into clans or coteries, each of which usually consists of a male, about three females and half-a-dozen juveniles; sometimes two males and about five females are present.

A coterie occupies a permanent territory of about 4,050sq m, which is covered by mounds of excavated soil and contains a complex of burrows with up to fifty entrances. It is handed down to succeeding generations. The members of a coterie show every sign of affection to one another, spending a great deal of time in grooming and kissing. When two dogs meet, they press their open mouths together. Since this greeting is only used by animals in their own territories, King suggests that an intruder regards the bared teeth as a threat and makes its retreat. Prairie dogs so enjoy kissing that they sometimes roll on the ground with mouths still in contact, then groom each other afterwards. No single age group has the monopoly of this pleasant pastime. Males and females kiss and groom each other, adults groom the pups, and the pups groom any animal they encounter. Pups have a delightful early life, for they are allowed to roam outside the territory, and are fondled and played with by neighbours and strangers alike. Whenever a youngster feels hungry, it only has to nuzzle under the nearest female to be given a meal. As the pup matures, it begins to be rebuffed by members of adjacent coteries, and eventually learns the boundaries of its own territory. During the breeding season, the coterie system breaks down. The females establish nesting territories of their own which they vigorously defend, and some males and juveniles move out to the suburbs where they dig new burrows for themselves. Each female produces a litter of about five pups each year. When these emerge from the burrows the coterie is re-established but there may be a continued emigration of adults.

Dominant males spend considerable time patrolling the territorial boundaries to prevent any take-over by another male. A trespasser is usually recognised by its hesitant manner. If it fails to retreat after the first rush, each of the rivals presents its anal glands for the other to sniff. This ritual continues until eventually one dog takes a nip at the rump of the other. The victor of the ensuing fight may give a two-syllable territorial call, delivered by the animal standing upright on its hind legs. Sometimes coteries are taken over by invading males, but fighting is extremely ritualised and rarely results in injury. A population in Kansas, which was studied by Professor Smith, showed little territorial behaviour. In this town the inhabitants gave a two-syllable bark when they were uneasy; other dogs, if also suspicious, joined in the alarm. Since the barks are delivered at a rate of about forty per minute, and may be kept up for over an hour, dog-towns can be very noisy. A different call is given when danger is imminent. A dog hearing this call doesn't wait to look around, but makes straight for its hole. Smith also describes an 'all clear' call, uttered when danger is past, or as a kind of greeting when dogs emerge on a fine, sunny morning. Sometimes this call is delivered with such gusto that the caller topples over backwards.

In its essentials, the coterie of the prairie dog is similar to the territorial system employed by captive house mice. Both ensure there is an even distribution of population. The centre of the colony offers greatest safety from predators, and it is important for the colony as a whole that this favoured site is occupied by breeding adults, the young and the old being forced to move out to the more vulnerable fringes as the population grows. The social organisation of the prairie dog seems ideal in many ways. The colonies are more or less permanent, for the larger ones date back longer than human memory can recall; consequently each pup is born into an optimum environment with established landmarks and traditions. It can grow up in reasonable security, and throughout its life of about twelve years, the animal is part of a community which insulates its members from many problems of the outside world. Although one observer found that each dog spent between a quarter and half its time keeping watch for predators, this still leaves ample time for feeding and enjoying life. With the exception of the carnivores, few mammals can afford to take time for play, but the prairie dogs, colonial marmots and ground squirrels, both young and old, indulge in all kinds of play activity. Arctic susliks have been seen chasing one another and even playing leap-frog.

Members of the marmot tribe have not penetrated to South America, in fact, only one genus of squirrel has reached northern Argentina. In the pampa, their niche is filled by a member of the chinchilla family, the viscacha, an amiable-looking animal, about the size of a domestic cat with black-and-white stripes on the face (page 143). The male's cheeks bear a moustache of long, very tough bristles (Fig 27). This seems to be the only instance of a rodent, showing a secondary sexual characteristic other than size. The function of the moustache is unknown. Viscachas live in colonies of between fifteen and thirty individuals, controlled by a single male.

Fig 27 *Viscacha, male and female*

W. H. Hudson recorded that in his time, the colonies were so close together that a man on horseback could see 100 from a single vantage point. They have now been cleared from many areas. Unlike most of the larger colonial rodents, the viscacha is nocturnal, leaving its burrow during evening to feed on vegetation. It has a strong instinct to clear the ground round the burrow, cutting down any tall plants such as maize, and dragging them with any other litter, to a pile outside the burrow entrance. The young Charles Darwin, while naturalist on the *Beagle*, observed that if anyone dropped a watch on the pampa, he would be sure to find it on a viscacha's rubbish dump.

Peasants used to visit the viscacheros to collect the litter heaps for fuel.

The viscacha should not be confused with the mountain viscacha which belongs to a different genus and is confined to the altiplano. Like the other rodents of this region, the mountain viscacha has soft fur, large ears and a long, furry tail. It is also diurnal. According to Oliver Pearson, who is one of the few naturalists to have studied these animals in the wild, they live in small colonies of between four and seventy-five individuals; each colony having its home in one of the rocky outcrops which dot the grassy slopes. During five month's observation, Pearson saw little fighting or evidence of attempts to preserve a territory. The animals preferred to cuddle up to each other as they basked in the sun. They seldom moved more than 80m from cover, and were very wary. Directly a man or dog approached, a whistling alarm note from a member of the colony caused the rest to 'freeze', or climb the rocks to get a better view; they only dived for the burrow if the intruder approached too close. A shorter note gave warning of hawks. Other mammals such as vicunas and leaf-eared mice also heeded the mountain viscacha's warning, and rushed for cover directly it was heard. Without the benefit of a communal security system, it is doubtful whether the species would have enough time to feed. Pearson described how blades of grass were eaten one at a time. Even when barley grains were provided, an animal spent twelve seconds over each grain and took three and a half hours to obtain its daily requirement.

Some sociable rodents are not only tolerant of members of their own kind, but also of members of other species. In Peru, guinea pigs often use the burrows of tucu-tucos, apparently without meeting objections from the owners. Here, the guinea pig is a commensal, and the benefits of the association are one-sided. In South Africa, an unusual form of co-existence has been established between a ground squirrel and a carnivore. The ground squirrel, *Xerus inauris*, sometimes called the bush meerkat, is common in many areas, especially in maize-growing districts. It lives in colonies of between three and five animals, in a burrow system with about twenty-five

Fig 28 *Mountain viscacha, an inhabitant of the altiplano of South America*

holes from which paths lead to feeding areas up to 875m away. The carnivore partner is the yellow mongoose, *Cynictis pencillata*, which is an inefficient burrower and uses the squirrel's holes. The mongoose also takes advantage of the colonial defence system, and rushes for a burrow directly it hears the squirrel's alarm-call. In return, the rodents have the services of a fierce fighter in defending the colony against rats, snakes and birds of prey. When live snakes were introduced experimentally into a colony, they were immediately chased out by the mongoose. Unfortunately, as so often happens when man appears on the scene, evolutionary adaptations which have contributed to survival over the years, become a positive threat. Since the mongoose was found to be an important carrier of rabies, gassing campaigns have resulted in the destruction of both members of this strange alliance.[177]

10

A WORLD OF ODOURS

Steeped as we are in fumes of tobacco, alcohol and diesel oil, it is fortunate that the sense of smell is not essential for man's well-being. Nevertheless it is the most esoteric of our five senses. Most of us can be 'turned on' by a whiff of some perfume or other, or involuntarily reminded of some long-forgotten event which took place earlier in our lives. Smell is the only sense in direct contact with the environment; hearing receptors are separated from it by the ear drum, and visual receptors of the eye are shielded by the cornea. Smell is a chemical sense like taste, but receivers in the nose are sensitive to minute quantities of airborn scent molecules; it is said that the human nose can detect musk at concentrations as low as 0·00004mg in a litre of air. The receivers consist of several million cells whose outer surfaces are covered with mucus; scent molecules are trapped by the mucus and stimulate sensory hairs which send impulses to the brain. Scent can only be perceived when air is moving through the nose. Very little is known about the volatile chemicals which give a substance its smell, or how olfactory cells are able to distinguish between the different scent molecules which fall on them; why some people are more sensitive to certain odours than others, and why sensitivity often varies according to time of day, sex and age. Although our scent receptors appear to be of little practical use, man's brain capacity owes much to the smell sense of his ancestors, for the cerebral hemispheres, the central computing system, has evolved from the olfactory centres of lower vertebrates.

An animal with the ability to perceive and recognise odours has several advantages over one which relies on sight and hearing. Eyes are useless in darkness, and need an uninterrupted field; sounds are particularly ephemeral and their audibility is influenced by wind and extraneous noises. Odour, however, is unaffected by light and noise, and has remarkable persistence. Scent molecules are often given off at a steady rate as long as the source persists. Muskone, for example, has been estimated to lose 10 per cent of its weight in 10,000,000 years, hence its value as a basis for perfumes. No matter how high the wind, once an animal's scent receptors are stimulated, it can orientate itself to catch more and more molecules until it reaches the source. The only limiting factor is that the odour substance must be exposed to air; scent molecules cannot be liberated if the material is covered by water.

One fundamental advantage of a keen sense of smell is in finding and identifying food. Many species of rodent are able to locate buried seeds—an ability well-known to those engaged in forestry reseeding schemes. The more odorous the seed, the more easily it is discovered. In a room of 18sq m, some American workers buried a variety of seeds beneath 5cm of peat. Deermice which were placed in the room found all the pine seeds, 85 per cent of the Douglas fir but only about 40 per cent of the wheat seeds. Similar results were obtained irrespective of whether the tests were carried out in subdued light or in total darkness.[72]

The ability to detect buried food is essential for species which habitually cover their food stores. It is also important that they should store only edible food and not waste time concealing inedible material or food which is decayed; any lack of discrimination could prove fatal during a spell of severe weather. In Britain, the most familiar hoarder is the grey squirrel, which is especially partial to hazel nuts. In an attempt to test a tame squirrel's powers of discrimination, H. G. Lloyd presented it with a series of pivoted glass tubes, some containing nuts but others with only cleaned brown pebbles. Within a few minutes the squirrel learned to tip the tubes and spill the contents, but it never moved a tube containing a pebble unless the stone had previously been placed in a heap of nuts for an hour. When the pet was older, it was given a mixture of sound nuts, maggoty nuts and empty shells. All the sound nuts were buried and all the empty shells refused, many without being handled. Empty nutshells weighted with lead were also ignored. The squirrel was less accurate in judging whether maggots were present since half the infested nuts were discarded and half were buried. This simple experiment demonstrated that a squirrel can accurately distinguish between the odour of sound and empty nuts. From the results of other tests carried out on grey squirrels in Ohio, it appears that squirrels vary a great deal in their ability to detect maggoty nuts, some recognising them by smell while others had to handle the nuts before rejecting them.

In human society, the accidental misappropriation of private property is a rare occurrence since virtually everything left in a public place bears a distinguishing mark. The rodent, however, faces the problem of discriminating between naturally fallen nuts, and those which have been collected by others. Possibly some pilfering does occur, but in the long term, the philosophy of survival of the laziest leads to extinction. Food already collected must therefore be recognised in some way. Illar Muul, while working on the food-storing behaviour of the American flying squirrel, *Glaucomys volans*, made an accidental discovery which explained how this species solves the problem. For his experiments, Muul needed large quantities of hickory nuts, and to save time, he attempted to use some nuts which had already been stored by squirrels. He noticed that if squirrels were given a choice, they chose new nuts in preference to those which had been used, at a ratio of 4:1. If the used nuts were washed in water, new nuts were still preferred, but only at a ratio of 2:1. After the nuts were washed in carbon tetrachloride and then rinsed in water, the squirrels could not distinguish between the two kinds. Before being stored, the nuts must have been marked with a substance which was soluble in tetrachloride but only partially soluble in water. By placing nuts just above and just below some caged flying squirrels, Muul found that the odour was deposited by the mouth, not the hands. Since this species often makes no attempt to conceal its nut stores, the application of a recognition mark is probably more important than to other species whose stores are well hidden. There is no evidence that an individual recognises its own foodstores, but providing every member of the community has contributed, there should be sufficient for all.

The marking of nuts by flying squirrels is a simple example of the utilisation of an animal's own body odour as a means of communication. Odours are used in this way by many classes of animals. Species of ant lay odour trails which guide workers to food stores, many insects attract a mate by odour, and injured earthworms liberate a warning substance which prevents others entering the danger-zone. There are countless other examples, but it is in the mammals that odours have reached their greatest complexity. Several different odour sources may be used: they are urine, faeces, special scent glands and possibly saliva. Scent glands have been described in fifteen of the nineteen mammalian orders, and about forty different types have been classified. Since rodents are easy to keep in captivity, far more is known about their odour signals than those of other mammals. Nevertheless the study of odours is still in the pioneering stage. Its students are in the position of children trying to operate a radio without having found the station selector or tuner. While an animal's response to the odours of another can readily be observed, we have little idea where the odour comes from, or whether a single odour can produce several different responses. Odour substances from related species may well be similar chemically, but there is evidence that they vary, not only according to species, but according to each individual. Experiments have shown that laboratory mice are able to discriminate between the odours of males and females of their own and other species, also between male mice of the same inbred strain.[16] My own striped hamsters, when sixteen days old, could distinguish their mother's urine from that of a strange female. In some species, this ability could assist a wanderer in finding its way back to the nest. Although the production of odour is inherited, its interpretation must sometimes be learned through experience, for it is possible to train mice to recognise the odours of commercial perfumes as signals. Female mice reared by perfumed parents, prefer a perfumed mate. It is probably unnecessary for members of a particular species to be able to recognise a great variety of odours, since each species normally keeps to its own niche, or perhaps shares it with one or two others. Odour signals may sometimes provide a cause of reproductive isolation, just as some birds which are virtually identical in appearance have different songs. They are by no means infallible. In the artificial habitat of the laboratory, the male black rat indiscriminately attacks brown rats of either sex, but the brown male sometimes tries to mate with a black male rat, although it attacks the females. It therefore seems likely that the odour signal of the male black rat resembles that of the brown female.[7]

It is important for a rodent to be able to recognise its own odour, for in many instances this is the only way the home burrow or territory can be distinguished. Many species habitually mark their territories as a means of getting to know them. When the cage of a pet mouse or hamster is due for cleaning, it is advisable to put some of the old litter back afterwards. If all traces of odour are removed, the animal will not settle down until it has marked all over the clean cage. Until it has been marked, any object placed in a rodent's cage or territory is generally regarded with suspicion. One captive agouti so mistrusted its sleeping-box after it had been cleaned, that it stayed out all night and died from exposure. Domestic rats leave odour trails, which in infested buildings are often visible as dark smears on the woodwork; in artificial colonies, such trails always attract the attention of other rats. The American deermouse is another trail follower; a fact which was confirmed by one of my captives which spent most of its life in total darkness. Its home was a kind of maze which consisted of twenty-one chambers interconnected by doors, which, when operated, flashed a numbered light on a panel outside. To travel from its nest to the feeding place, the deermouse had to take a circuitous route of about 2m and open nine doors. This journey could be accomplished in three seconds. After a time, faint brown smears appeared on the plaster of Paris squares forming the floor of the maze, and a distinct trail could be seen leading directly from the nest to the feeding-place. When the plaster squares were transposed, the released mouse still followed the odour trail which now led in a different direction, taking seven seconds to reach the new, foodless terminus. When the bases were moved to disrupt the trail completely, the mouse took two and a half minutes to reach the usual feeding-point.

Some rodents leave an odour mark in special places, in much the same way as neighbourhood dogs use lamp-posts. Grey squirrels gnaw patches of bark on growing trees then urinate over the injury, in time creating a dark stain. Many marking points are on the undersides of branches—situations protected from rain, but necessitating any callers to urinate while upside-down. On smooth-barked trees such as beech, marking points are between root buttresses, or on a root just breaking the soil surface. When the tree cannot replace the bark, the wood is exposed permanently, and sometimes a knob of callus tissue forms where marking has been carried out by generations of squirrels. Both sexes take part in the marking ritual.[157] There is no definite evidence that the faeces of rodents provide chemical signals, but it was only in 1958 that Mykytowycz discovered the significance of rabbit droppings as a means of social communication. He found that rabbit dung-hills serve a territorial function, and are made up of especially odorous pellets; the amount of odour being directly related to the rabbit's position in the social hierarchy. The faeces acquire the odour through a secretion of the anal glands. These glands are well-developed in dominant individuals. Upon encountering a heap of pellets bearing a strange odour, a rabbit's reaction is to drop a scented sample of its own. Unmarked faeces are just scattered anywhere, only those marked with secretion are placed on the familiar heaps. Mountain viscachas also deposit their droppings in heaps, often at special places. Some sites are used so often that they cover large areas of ground.[117] The African pouched rat, *Cricetomys gambianus*, sometimes uses an outside latrine which could serve a territorial or social function; captives have been seen to stand on their hands in order to deposit faeces high on the wires of their cages.[50] Mykytowycz, during his work with rabbits, noticed that females, after sealing a stop containing young, deposited a few faeces and a little urine which apparently served to warn off intruders. The faeces of striped hamsters have an inhibiting effect on the movements of strangers. If hamsters are given the choice of two passages in an experimental cage, they are reluctant to enter the passage with a few strange faeces at the entrance. In some species, urine and glandular secretions have an attractive effect, depending on the sexes of donor and recipient. Grasshopper mice, when presented with balls of cotton in which others had nested, always showed the most interest in those which had been used by members of the opposite sex.[134] On the other hand, male gerbils are attracted to sebum from other males, but not to sebum from females.[159] Odour preferences of this nature may influence the results of live-trapping experiments; female house mice, for instance, show a marked preference for traps previously used to accommodate males.[132]

The males of several 'porcupine-like' rodents do not stop short at marking their territories, but also mark their intended mates by sprinkling them with urine. This process, which is termed enurination, has been observed in the green acouchi, guinea pig, mara, paca, and American porcupine. When a male porcupine is introduced into the cage of a female, it urinates on one of its front paws, probably to transfer its own mark to the cage. If amorously inclined, it makes a quiet whining song and approaches the female, rearing on its hind legs after sniffing her body. Should the female be receptive to such advances, she also rears up and the pair rub noses. Then, decorum is abandoned when the male urinates with great force, thoroughly drenching his prospective mate and splashing drops of urine to a distance of over 2m.[139] The performance is hardly comparable to the human kiss, since the female shows no sign of enjoyment and sometimes runs away. It has been suggested that this marking of the female serves as a kind of engagement ring in warning off rival suitors.

All the species of rodents examined have scent glands of some kind. Some have different types, which presumably serve several functions. Unlike chemical signals involving urine marking, glandular secretions do not result in a significant loss of fluid—an important consideration for desert-dwellers. The positions of the scent glands vary according to an animal's mode of life; the most usual are the sides of the face, the flanks, and the under-surface of the body. Glands at the sides of the lips, similar to those used by the flying squirrel for marking nuts, are well developed in American ground squirrels, of the genus *Spermophilus*, and are present in many 'rat-like' rodents. Facial glands have been described in the American porcupine and African ground squirrel, *Xerus erythropus*. Marmots have been observed using their facial glands to mark objects in their territories. Richardson's ground squirrels have been seen rubbing the sides of the head on the ground around the burrow entrances, and it seems that an odour gland is positioned just behind the ear; during mutual grooming by members of a colony, it is this part of the head which is given greatest attention, so in this species the odour may serve a social as well as territorial function.[121] Glands on the flanks are particularly useful for marking the walls of burrows, and they are well developed in hamsters and the European water vole. By scratching the flanks with the hind feet, the secretion can also be transferred to the ground. When the water vole is agitated, it sometimes drums with its hind feet, no doubt reinforcing the sound signal by liberating odour at the same time; even three-week-old water voles have large glands and are able to take and hold territories at this early age.[155] Jerboas and kangaroo rats frequently sandbathe; the habit is said to be essential for the health of kangaroo rats, for if captives are prevented from sandbathing the fur becomes matted by secretion from the gland on the back, and sores may develop. Favoured sandbathing spots may act like the urine sign-posts of squirrels, for kangaroo rats show a strong inclination to investigate the sites used by other individuals.[45]

Glands opening on the underside of the body are well placed to allow a low-bodied animal smear secretion as it runs along. Anal glands are present in the grey squirrel, edible dormouse, guinea pig and marmot. The anal region is simply pressed on the ground or the object to be marked. In many 'rat-like' rodents, the part of the body nearest the ground is the prepuce, the sleeve of skin covering the penis. Many species possess a paired odour gland which opens by a duct at this point. In the house mouse, it is the secretion of these glands which give the distinctive mousy odour. In this species, and in domestic rats and musk rats, the glands are present in both sexes, but in others they are characteristic of the males. Another useful glandular site is the midline of the chest or abdomen. Mid-ventral glands have been discovered in several species, including woodrats, deermice, hamsters, cotton rats and gerbils. In the Mongolian gerbil, the hairs surrounding the gland are specialised for depositing secretion, being grooved and pointing backwards. Mid-ventral gland secretion, at least in gerbils and hamsters, does not smell unpleasant but resembles that of burnt feathers. Male European wood mice (*Apodemus*) have a glandular area on the underside of the tail. When squeezed, a milky secretion is exuded. In England, these glands are present in both the yellow-necked woodmouse, *A. flavicollis*, and the common woodmouse, *A. sylvaticus*, but in Germany they occur only in the former species.[56] The Australian kangaroo mouse (*Notomys*) has a glandular pocket in the throat.

Fear is an emotion which even man cannot conceal. It affects the body in various ways, including loss of bladder control, excessive secretion by skin glands, and the hair stands on end. These reflexes, which are merely embarrassing to man, provide an important defence mechanism for many mammals. Few observations have been made on rodents using odour for defence. I have noticed it on only two occasions. After one of my deermice had been alone for several weeks, a stranger was quietly introduced into its large cage. The newcomer remained motionless and was unnoticed by the original occupant which strolled unconcernedly along its usual path. When it suddenly came face to face with the strange mouse, it 'froze' and liberated an acrid odour which I could detect over a metre away. The second occasion was when I startled a captive water vole by

Fig 29 *African crested rat*, Lophiomys

suddenly opening the door of its pen, this time the odour was almost stifling at a distance of over 2m. Similar reactions have been described in the South American burrowing mouse, *Oxymycterus rufus*,[5] and in the East African crested rat, *Lophiomys imhausi*. The crested rat (Fig 29) is a large forest species which looks more like a badger than a rodent, since its long fur has a pattern of dark and light, greyish stripes. When alarmed, it erects the mane on its back to expose lateral patches of short brown hairs extending from the neck to behind each shoulder. These hairs are glandular and have a remarkable structure (Fig 30). Each is a hollow cylinder with a finely pointed tip, but instead of its wall being solid, it consists of an open meshwork, through which, presumably, the odorous substance is liberated. The real purpose of these glandular hairs is unknown but they may discourage predators. Being a slow-moving species, the crested rat appears to need special protective devices. It is possible that the curiously strengthened skull gives shelter against attack by snakes. The only other rodent with an armour-plated skull is the paca of the South American forests.

The liberation of a 'fear' odour can only be useful if it can be interpreted as an alarm by other members of the species which are in the vicinity. It has been shown that laboratory rats can distinguish between the odours of normal rats and those produced by rats which have been given electric shocks. Rats were trained to press a bar when a current of air was passed through their chamber. If the air had previously passed over normal rats, bar-pressing was rewarded by some sugar solution being dropped into the cage; if the air had come from shocked rats, the bar-presser himself received a shock. The rats soon learned to sniff at the odour source directly the air supply was introduced; if the odour came from shocked animals, they stopped pressing and sometimes retreated to the far end of their chamber.[164] It remains to be seen whether the odour produced by a shocked rat is a kind of 'fear' odour, or indeed whether it has an inhibiting effect on others if it is not painfully reinforced.

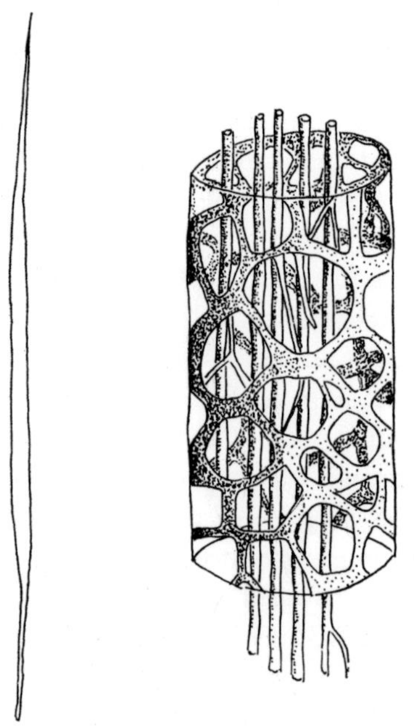

Fig 30 *Glandular hair of crested rat; part is greatly magnified to show structure*

The odour glands of one species, far from assisting in survival, once contributed to its near-extinction. For hundreds of years, beavers have been persecuted for both their fur and the castoreum which was a valuable ingredient of perfumes and medicines. Castoreum is an extremely complex substance, and over forty constituent compounds have been identified. It is produced by paired, pouch-like glands which open into the cloaca of both sexes. Fortunately for the beaver, the demand for castoreum is now met by the more mundane coal-tar derivatives. In the hey-day of the castoreum trade, there was a widespread belief that the substance was produced by the beavers' testes, and the castoreum hunters, aware of the internal position of these organs, probably fostered the belief to dissuade others from entering the lucrative profession. Eugenius Philalethes, in his *Brief Natural History* of 1669, was not taken in, and discounted the popular belief that,

> . . . the beaver, being hunted and in danger to be taken, biteth off his stones, knowing that for them his life only is sought, and so often escapeth; hence some have derived his name Castor, a *castrondo seipsum*, also who likewise affirm that when the beast is hunted, having formerly bitten off his stones, he standeth upright, and sheweth the hunters that he hath none for them, and therefore his death cannot profit them, by means whereof they are averted and seek for another.

The beaver uses the castoreum for marking its territory, but other beavers are not repelled and deposit their own marks in the same place. It has long been known that beavers can be lured into traps baited with castoreum. The most usual communication posts are on bare mounds or alongside a trail, but in some regions the beaver makes little islands in streams for this purpose.

For several decades it has been known that certain animal odours produce involuntary reactions in the animal which smells them. The actions of such substances are similar to those produced by hormones, the chemical messengers of the blood. Odours which function in this way are termed pheromones. For possibly more than a century, insect collectors have attracted male moths by placing a female of the desired species in a gauze cage or tethering her with a piece of thread—a process known as 'assembling', but it was not until 1955 that any attention was given to the likelihood of pheromones occurring in mammals.

Beginning with the work of the Dutch endocrinologists, S. van der Lee and J. M. Boot, several remarkable discoveries have been made on the ways in which the reproductive patterns of mice are influenced by the odours of other mice. These two workers found that when several female mice were caged together, their oestrus cycles became highly irregular, often stopping altogether for long periods. The cycles returned to normal when the mice were placed in separate cages or had their olfactory bulbs removed. In the same year, W. K. Whitten of the Australian National University, reported that the odour of male mice can initiate the oestrus cycles of females. Whitten discovered that his mice mated more frequently on the first night after pairing if the male had been kept in a small basket in the female's cage beforehand; the previous presence of the male was unnecessary provided it had contaminated the female's cage. Some odour substance from the male had obviously modified the female's oestrus cycle in some way. Subsequently, it was found that oestrus could be induced by placing a drop of male urine on the nostrils of a female mouse.[173] The pheromone responsible is not produced by any special accessory odour glands since the effects are obtained from urine collected straight from the bladder of a male. In a more elaborate experiment devised by other workers, a cage of male mice was placed so their urine would fall into a cage of females below, and at the same time, by a kind of wind tunnel arrangement, other cages of females were placed both upwind and downwind of the males. It was found that in the cage placed 2m downwind, the proportion of females which came into oestrus was the same as that of the group exposed to male urine. Few females in the upwind cage came into oestrus.[172] Since urine from castrated males has no effect on the female reproductive cycle, the pheromone is obviously linked to male sex hormone, and brings the female into heat when a prospective mate is around. The bizarre urine-dowsing ritual of courting porcupines probably serves a similar function. Such a sex stimulant must be especially advantageous among the more solitary species, for males cannot spend too much time casting around for a female in receptive mood. The minute quantities of pheromone prepare the way just as efficiently as the lengthy and exhausting courtship rituals practised by birds and man.

In 1959, an English scientist, H. M. Bruce, discovered another interesting property of male mouse urine. If a recently mated female was removed from the company of her mate and placed with a strange male, pregnancy was usually blocked, and the female returned to oestrus. At first, there was some doubt whether the effect was due to the sight or smell of the strange male, but with a restraint sometimes rare in researchers, Bruce felt 'that the conclusive experiment of blinding female mice was not justified'.[115] Instead, the females were exposed to urine from strange males. The results were identical. More recent studies have shown that the strange-male odour is less effective as a pregnancy block if the stud male is removed very quickly after mating; it seems as if the female needs several hours to become sensitised to her mate's odour.[91] The urine of the male mouse must contain at least two quite different pheromones because the pregnancy block is produced by urine from castrated as well as normal males. The 'Bruce effect' has been demonstrated in the laboratory mouse, deermouse, and the European vole, *Microtus agrestis*.[30]

One can only guess at the evolutionary significance of the pregnancy-block pheromone. In a closely knit social group of rodents, a female is likely to mate with her own adult offspring or close relatives. If a strange male, however, is able to replace incestuously acquired genes with his own, the amount of inbreeding is reduced. It is difficult to suggest the purpose of the oestrus-inhibiting pheromone produced by crowded females. Possibly it serves as a birth-control mechanism in over-populated situations.

The pheromones in urine are not solely concerned with reproductive cycles. Exposure of young mice to male urine, accelerates their growth and maturation, while exposure to urine from virgin females is associated with a slow growth rate.[33] Male urine also increases the activity of adult males, irrespective of whether it is fresh or fermented. As urine from castrated males does not have this effect, a sex pheromone seems to be involved.[129]

Whatever the real purpose of rodent pheromones, their properties should be borne in mind by anyone trying to breed rodents for classroom study or as pets. The chances of success may be slight if the females are kept in close proximity to each other, even if in separate cages. Several species such as deermice, steppe lemmings and striped hamsters sometimes breed most readily if two males are kept with each female; particularly aggressive females must, of course, be kept separate, but cages of males should be nearby.

The lateral scent organs of the water vole, preputial glands of the rat, and the mid-ventral glands of the Mongolian gerbil and deermouse appear to be under the influence of sex hormones, becoming active at sexual maturity. The male water vole has larger glands than the female, and the amount of secretory tissue is related to the weight of the testes.[155] Male gerbils mark much more frequently than females and their glands are twice as large; if the male is castrated, the gland shrinks but can be restored by injections of testosterone.[158] The ventral gland is not, however, essential for successful breeding, since litters have been sired by males whose glands had been removed by surgery.[101] Since the size of the odour gland is related to the degree of maleness, the intensity of its signal may well indicate social position. If, for instance, a gerbil is placed in the territory of another, its marking is usually inhibited by the presence of the owner's marks; but if the intruder is a particularly aggressive individual, it marks the usurped territory even more vigorously than its own.[159] The possession of a ventral gland by the female gerbil may be due to male hormones being produced by the ovaries, because if these organs are removed, the injection of oestrogen prevents the gland developing.[42] Odour glands of this type prevent mistaken attacks on females, or on the young and old whose odours are too weak to induce an aggressive response. The lateral glands of hamsters seem to serve a similar function to the gerbil's ventral gland. One experimenter allowed pairs of male hamsters to fight in neutral territory, then placed each combatant in the empty cage of its opponent. In every test the winner of the fight marked more frequently than the loser. Female hamsters, as if dismissing the males' marks as of little consequence, mark more frequently in empty female cages than they do in cages formerly occupied by males. Male hamsters mark very little when put in a cage which has housed a female on heat, the reduced marking is believed to be due to the inhibiting effect of a special vaginal secretion.[123] If the female is present, she examines the male's lateral glands, becomes excited by their odour, and the resulting secretion from her clitoral glands serve as a green light to the male's advances.[108]

Among all the species which have been studied, male rodents rarely attack females, their odour apparently inhibiting any aggressive tendencies. If male mice are rubbed with urine from adult females, aggression by other males is often prevented.[41] The females of some solitary species such as hamsters, are notably belligerent towards males, biting generally their most vulnerable parts. In the wild, the male can probably escape without difficulty, but it would seem that just as a female stimulating odour has survival value for some species, a female tranquillising odour would also have its uses. The large mid-ventral gland of the male striped hamster may serve this function. When a male is persistently attacked by a female, it eventually rolls on its back to expose its gland, the female then turns her attentions from the male's posterior, and after muzzling the gland, quietens down and walks away.

Although our knowledge of rodent odours is based on the study of a few species kept in captivity, it is obvious that odours play an important part in the lives of many species. It is equally certain that it will be many years before the full potentials of those volatile chemicals are known. As for perfumes used by women, one French researcher pacified aggressive mice by spraying them with a scent by Dior.[130]

11

THE HOME

It is doubtful whether any mammal wanders at random over the earth's surface. Even the migratory caribou and nomadic human tribes generally follow familiar and well-tried routes. Most small mammals are quite restricted in their movements, each individual confining its activities to a limited area which is known as the home range. Here, the animal can learn to exploit all the available feeding- and hiding-places, and just as a motorist avoids the police patrols of his home town, it is fairly secure from predators. All the species of rodents studied seem to have a home range, and only leave it under exceptional circumstances, such as the population upheavals mentioned earlier. The size of the home range is difficult to estimate, since few rodents are conspicuous enough to be located by direct observation. The most usual method of study consists of trapping an animal alive, marking it and then releasing and trying to recapture the same specimen in a series of live-traps arranged in a grid pattern. Sometimes the use of baited traps gives a false indication of the range since some individuals habitually visit the nearest trap, willingly tolerating the slight inconvenience of a few hours incarceration in return for a free meal. During Professor Smith's studies on prairie dogs in Kansas, some individuals so delighted in visiting his traps that they had to be released three times a day throughout the trapping period. Errors of this kind occur only in methods which rely on recapture, so other techniques are sometimes used. A few

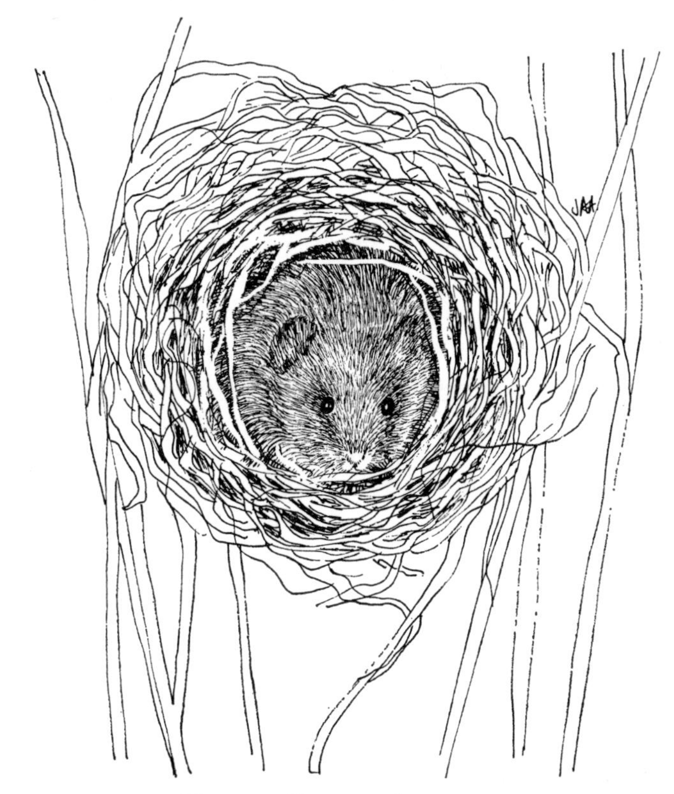

Fig 31 *European harvest mouse*

species such as chipmunks and squirrels can be dyed and their movements followed by direct observation. Mice and voles are sometimes traced by their footprints; in the absence of snow or mud, boards sensitised with lamp-black or some other substance are placed in the study area and examined at regular intervals. Unfortunately for the animals concerned, the recognition of individual tracks is only possible after one or more toes have been amputated. Another technique involves marking the animals with radio-active material before release, then locating them with the aid of a Geiger–Muller counter. With some species, if the location of the nest or food store is known, specially marked seeds can be distributed at known locations, then, with luck, recovered from the animal's food store. The wood rat's penchant for collecting shiny objects has been exploited by observers eager to trace its movements. The technique involves simply distributing balls of tinfoil in trees and other places, then recovering them from the nests.

All methods of estimating home range have their limitations, and are extremely time-consuming. Consequently, most investigations have been carried out on North American and European rats and mice, in areas readily accessible to students. In the small number of species studied, the size of the home range was generally between 1,000 and 16,000sq m. To a great extent of course, the size is influenced by the amount of food and cover. For example, a species of deermouse was found to have a larger range in the open mesquite of northern Mexico than in a richly vegetated part of Michigan. Studies made in New Mexico revealed that male grasshopper mice occupied ranges as large as 23,000sq m. Males usually have larger home ranges than females. A detailed study on bank voles, carried out near Oxford, England, showed that range size varied between 2,116 and 523sq m, but the average range size of the males was 1,515, and that of the females, 1,120sq m.[81] Some species have a very small home range. In some buildings, female house mice had a range of 50, and males of 112sq m.[44] The size of the home range is not constant but varies according to season, changes in population density, and other factors. Some may be roughly circular in shape, some, including those of water voles, may be long and narrow, while others may be quite irregular for no readily apparent reason. It seems likely that some species of rodents, like certain birds, do not find their way just by memorising landmarks. Some quite remarkable feats of homing have been recorded. There have been instances of European wood mice and bank voles returning 800m to the place of capture, and a yellow pine chipmunk, *Eutamias amoenus*, has returned from a distance of 1·6km.[21]

Understandably, there is often confusion between the terms 'home range' and 'territory'. Strictly speaking, a territory is that part of an animal's range which is defended against intrusion by members of the same species. Since the concept of territory was first formulated by H. E. Howard in 1920, great strides have been made in the study of territorial behaviour in birds. Because bird behaviour revolves largely around colour and song, its study has been comparatively simple; even stuffed specimens and tape recordings elicit attack by birds defending their territories. Wild rodents are far more difficult to study. Few observations on their territorial behaviour have been made, and the term 'territory' is often used rather loosely. In several species the defended area is small; the wood rat defends only its house, and the hamster only its burrow and immediate surroundings.

It is possible that other parts of the home range are defended, especially if marked by the occupant's odour, but there is only a slight chance of an encounter between resident and trespasser being witnessed. Some squirrels apparently defend the eating- and nesting-places, and nursing pine chipmunks are said to be particularly aggressive near the den. One observer has described how a male tassel-eared squirrel defended a 20m Douglas fir against seven intruding squirrels. It chased them out of the tree, then barked continuously and tapped the trunk with its front foot; in this instance, the spirited defence of the tree was due to the presence of a female squirrel at the top.[126]

In spite of the defence mechanisms described in another chapter, most rodents face the constant threat of attack by predators whenever they expose themselves. The majority need a reasonably secure place in which to retire after feeding forays, as well as to rear their young. There are a few exceptions; the prehensile-tailed porcupine often sleeps in the branches of a tree, and the Bahaman hutia does not need even a shelter for its young since they are able to move about soon after birth. Many rodents, however, make some kind of nest which serves a number of important functions. It provides protection from enemies, shelter from the elements, security for the young, and, in many cases, allows the occupant a degree of privacy from its fellows. Small species often have difficulty in keeping warm, and the insulating properties of the nest are especially important for those whose young cannot generate sufficient body heat until they are several days old. As a rule, the most substantial nests are made by pregnant females. The amount of nesting material used depends largely on external conditions. At temperatures around freezing point, house mice were found to make nests weighing about 15g, but at 30° C (86° F) they made no attempt to build. Similarly, when laboratory rats were provided with strips of paper, the weight of paper used for nest-building was influenced by the temperature.

The nests of rodents show far less variety than those of birds. They are usually roughly spherical; those of some species have one entrance hole, others two, but often the occupant simply pushes through the wall and closes the hole behind it. Almost any convenient material may be used for building the nest; in the absence of grass or leaves, paper or polythene may be shredded to make an adequate structure. One mouse of my acquaintance created grave catering problems when it escaped into a well-stocked larder and stripped all the labels from food cans to make its nest in a pudding basin. Exposed nests, such as those of squirrels, usually consist of a firm outer layer of twigs or coarse vegetation and an inner lining of finer material. Few observations have been made on methods of nest construction. First, the animal has to collect material and carry it in its mouth to the site. Some ground squirrels have a most efficient way of carrying grasses. They are held lengthwise in the mouth, then folded into neat packages by brushing the hands along the sides of the face. In many species, including hamsters and voles, the builder just sits in the middle of its pile of nest material and with its mouth keeps pulling up the edges of its platform until a roof is formed over its head. The European harvest mouse, which makes a very neat spherical nest, sometimes tears grass for its construction by sitting across a blade, biting into the middle and ripping out the centre. The European dormouse often uses shredded honeysuckle bark for making its nest; a habit which sometimes betrays the animal's presence to the observant naturalist. Both species build just above the ground. In many districts, the harvest mouse favours juniper bushes whose dense, prickly leaves provide additional protection.

Many climbing species use tree-holes as nesting sites. The forest sicista mouse of Europe gnaws out twisted channels in rotten trunks; it sometimes uses old woodpecker holes and has even been known to share a hole while a woodpecker was in residence. The pencil-tailed tree mouse, *Chiropodomys gliroides*, of the Far East, frequently nests inside a bamboo. After making an entrance hole and getting inside, it gnaws through the internodal septum and makes its nest in the internode either above or below the entrance—a stratagem which is likely to confuse predatory snakes, and restrict draughts.[100] Many species use birds' nests occasionally. In Africa, the climbing rat, *Thamnomys dolichurus*, has been known to rear its young in the suspended nest of a golden weaver; European dormice sometimes use wrens' nests, and in the salt marshes of Georgia, the rice rat modifies the nests of the marsh wren. In all parts of the world, tree squirrels frequently utilise nests of crows and other large birds. In some parts of South America, marsh rats (*Holochilus*) achieve maximum security by weaving their nests on reeds some 30cm above the water; when disturbed, the rats are said to leap from their nests and swim away.

Being so adaptable, small species generally have little difficulty in finding a suitably sheltered site for the nest. Where conventional sites are scarce, unorthodox places are used. The African pigmy climbing mouse (*Dendromus*) sometimes nests in a fallen corn cob, and the Javanese flying squirrel builds in coconuts which have been gnawed open and eaten by other rodents. Large species, unless they are able to burrow, face far greater problems in finding adequate shelter, and the distribution of some species may be influenced by the number of sizeable holes and crevices available. In Africa, the crested porcupine is most common in areas where there are plenty of rock crevices, termite hills or buttressed tree roots in which it can make a den. The North American porcupine shelters in tree holes and hollow logs; in some districts there is such a demand for suitable 'den' trees that they may be used by a succession of individuals for many years. The mountain viscacha and chinchilla of the Andes are also poor burrowers, and their colonies are only found in rocky areas.

A few species have become independent of natural shelters through being able to construct their own. Such structures are more elaborate than the usual rodent nest, and normally consist of several chambers under a single waterproof roof. Since these species are capable of burrowing, the origin of the habit of house-building is conjectural, but because of it, they are able to colonise areas where burrowing is impossible and natural shelters non-existent. The house-building ability has proved especially advantageous to the American wood rats (*Neotoma*) which are distributed over most of the USA, British Columbia, Mexico, Nicaragua and Guatemala. Besides inhabiting the more agreeable types of habitat, some of the twenty or so species live in deserts, some in rocky areas and others in swamps. The wood rat is very ordinary in appearance, rather like a large brown rat but with a soft coat and furry tail, yet it can build houses up to 2·4m high and 3m in diameter. The names pack rat and trade rat are more appropriate than wood rat, for the animal is a great collector, not only of food and building materials, but of glittering oddments. If, during its nocturnal wandering, something bright catches its eye, it drops whatever it is carrying and picks up the more attractive item; many a camper has been disconcerted to wake up and find one of his spoons taken in exchange for a stick, or lump of mud. The collecting mania still persists when the wood rat takes up residence in a human dwelling; the nightly din it creates by dragging its acquisitions over the bedroom ceiling rapidly leads to the eviction of either man or rat. According to an Indian legend, the wood rat is the descendant of a giant rat which, at one time, crept into villages to carry off Indian children, or steal valuables to store in its cave. Eventually the monster was caught and thrown over a precipice, but during its descent it shrunk in size like a punctured balloon, landing at the bottom as the small harmless rat we know today, but still retaining its old thieving tendencies.[119]

Most of our knowledge about wood rats is due to Jean Linsdale and Lloyd Tevis who, for ten years, studied the dusky-footed wood rat, *Neotoma fuscipes*, in a reserve on the Californian coast. Most wood rat houses in this area are built of sticks piled around a tree trunk or branch which provides anchorage. Each house has several entrances, at least one nesting chamber containing a nest of shredded bark or thistledown, a foodstore, and a latrine. Usually there is also a sheltered feeding-terrace with a wide view; and below each house, a short, cul-de-sac tunnel probably serves as a retreat in case of emergency. If a home is to keep its occupant dry, particular attention must be paid to the maintenance of roofs and drains. The best wood rat houses have steeply pitched roofs which are kept in good repair during the winter. Quite large pieces of building material may be used; one rat was seen hauling a stick 10cm in diameter and 400cm long, on to the roof of its house. Preferred housing sites are near the top of a slope where there is good natural drainage; when the house is on sandy soil, holes dug in the floor help to keep it dry. In some areas there is a great demand for houses and directly one becomes vacant, a rat moves in. Should a house be left empty for some time, it seems to be no longer recognised as a house, possibly because it lacks the odour of its former occupant. A homeless rat, however, being vulnerable both to predators and weather, cannot afford to be fastidious in choice of home, and may use an old derelict house as a temporary shelter while a new one is being built. Good houses in suitable sites may be used by generations of rats, and since each occupant carries out repairs and extensions, old houses can be very complex. Linsdale describes one which contained four nests, seven latrines, and various passages arranged on three levels. As in old human dwellings, sanitation seems to be a perpetual problem and leads to a proliferation of latrines. One meticulous observer reported that the wood rat produces 124 faecal pellets a day, or 3·25gal a year. In the 1920s, wood rat faeces were used as a garden fertiliser in parts of California, most houses yielding up to three sackfuls while others filled ten or even twenty sacks.

The desert wood rat, *Neotoma lepida*, which lives in North American deserts, makes different styles of house according to the sites and materials available. Observations made at several localities in the Mohave desert revealed that in rocky areas the rats seldom made a house, but used rock crevices, blocking the entrance with piles of twigs or rocks. Rats living in the creosote bush scrub also used crevices but often made houses of acacia twigs, the recurved spines helping to consolidate the structure and possibly serving as a deterrent to predators. In one area where no other material was available, joints of cholla cacti were used as building blocks, being piled around small desert shrubs which acted as central supports (Fig 32). Although they adhere to each other by means of the spines,

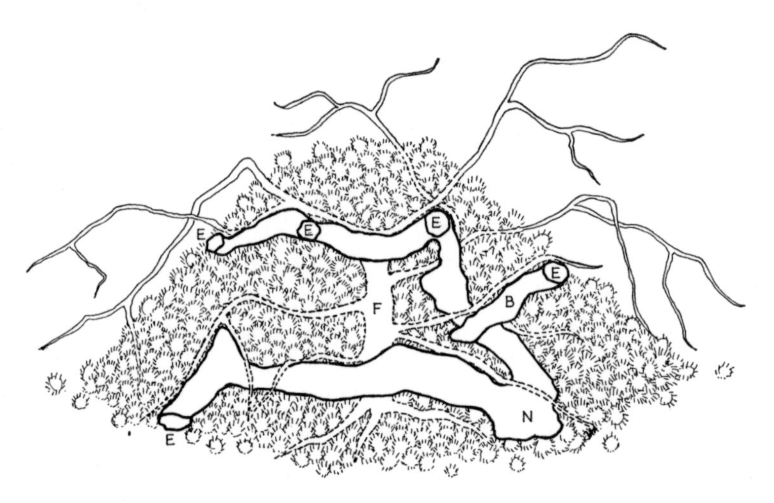

Fig 32 *Cactus house of wood rat,* Neotoma lepida*:* (N) *nest-chamber;* (E) *entrances;* (B) *blind passageway;* (F) *creosote bush acting as support*

cactus joints are not the easiest of building materials to use. In the intense heat the watery tissues soon decompose and the blocks collapse, forcing the tenant to resurface the house continually with fresh cholla joints. Naturally enough, wood rats treat the cactus with respect; when a joint is to be carried home, it is gingerly manipulated and a single spine is grasped between the incisors. Although a house which is fortified like a barbed wire entanglement is unlikely to encourage trespassers, the occupant can enjoy little comfort, and there have been instances of wood rats becoming entangled with the joints and receiving fatal injuries.[25]

The wood rat's house-building ability is rivalled by members of an unrelated Australian genus, aptly known as stick nest rats (*Leporillus*). When their houses were first discovered in the plains of the Murray-Darling by a survey party in 1838, they were believed to be made by aborigines for signal fires. Up to 2m in diameter and over 1m in height, the houses are remarkably like those of the wood rat. They usually have about five entrances around the base, passages connecting with the soft interior nests, and also a burrow in the soil. The sticks, which are sometimes 1m long, are generally placed around a stunted bush, but if this is impossible, small stones are put on top to prevent the roof from being blown away.[162] Some houses are built over rabbit warrens which provide ready-made escape routes. In desert regions, rodent houses of this type may be the only shelter for miles. Consequently they are often shared by other animals. Those of the stick-nest rat often house bandicoot and penguins, and Linsdale found a family of wood rats sharing their accommodation with a shrew, alligator lizard and two tree toads.

The master builder of the rodent world is the beaver, and its house so resembles the Indian lodge that it has been given the same name. Varying in size, lodges reach a height of over 2m and a diameter of 8m; constructed of sticks and plastered with mud except for an air vent at the top, it is capable of supporting the weight of several men, and cannot be broken into without an axe. It has already been remarked how the beaver's building ability is due to natural selection rather than foresight, and the provision of a ventilation grid in the roof seems to be no exception. When the animal climbs up the outside of the lodge while carrying armfuls of mud from the banks, it is exposed to birds of prey. Wilsson suggests that the risks involved in plastering the extreme top far outweigh any advantages, and as a result of natural selection, the top is left unfinished and permeable to air. The interior of the lodge is usually smeared with mud, and the top covered with moist sticks which maintain a low temperature and high humidity in summer. In winter, insulation is increased by a covering of snow. Temperature readings made at a lodge in Ontario, showed that in January and February the inside temperatures were always just above freezing point, varying by less than one degree centigrade, while outside the daily average temperatures fluctuated between – 21° C (21·2° F) and 6·8° C (44·2° F).[153] Lodges may endure for a long time; Ognev mentions one in Russia which was used regularly for forty years.

Some lodges are made on firm foundations; others, built on piles of wood placed in the water, often become islands which float in deep water. The beaver only builds lodges on marshy land. In drier areas it digs burrows in the banks of streams. The entrance, as with the lodge, is always through a short sloping tunnel or plunge hole which comes out near the bed of the stream or lake. The floor of the burrow is covered with chips of wood and twigs, and sometimes the accumulation of food material leaves little room for the occupant. When this happens, soil is sometimes dug from the ceiling to make more headroom, and as a result, the whole excavation may collapse through heavy rain or animals walking over the top. Some authorities believe that the lodge evolved through the beavers' attempts to repair holes in the burrow roof. As well as lodges and burrows, beavers also use temporary dens in summer, sometimes beneath the roots of alders or in dense vegetation; during a flood, one nest was found 3m above water in a willow tree.[114]

In marshes, the homes of both the muskrat and round-tailed muskrat consist simply of mounds of vegetation with a nest near the centre. Like the beaver's lodges, they are almost invariably provided with one or more underwater exits. The muskrat also lives in burrows in river banks, with some entrances below water level and others on land. Land entrances are plugged with vegetation and serve as air vents. Should the water level fall, exposed holes are plugged in a similar way and lower ones are used.[169] A water-filled entrance tunnel is a useful adjunct to the home of an aquatic species; besides keeping out cold air and predators, it still allows the occupant to move in and out even when the water outside is frozen; the beaver's plunge-hole also serves as an indoor pool in which the kits can practise in safety. Underwater entrances are often used by the European water vole and the brown rat. The African water rat seems to have no permanent nest or burrow, but during the breeding season it uses a combination of both to form what must be one of the most elaborate structures used by any rodent. Every two months, for about a year, I camped on the shores of a Malawi swamp without seeing a trace of this rat, apart from a few skulls which regularly appeared in the pellets of the owls living in a nearby baobab. Early one June morning, as I was visiting the trap-line, an unmistakable dark form ran across the path and vanished inside a grass nest which lay on the ground. The nest was empty by the time I reached it, but a few minutes later another rat ran into a similar nest. This time, a hat was tossed over the top to prevent any escape. When this was also found to be empty, it was closely examined. The 200mm ball of grass consisted of two compartments, one on the surface and the other in a shallow excavation immediately below; they were so arranged that a predator could search the surface nest without realising the existence of a basement. When the complete nest was lifted from the shallow hole, it was seen that a tunnel extended through the soil in the direction of the swamp. Subsequently, several other nests were found and when their tunnels were excavated, they were found to be identical, being just over 1m in length and with the blind end filled with water. Since only two juveniles were recovered from one burrow, the occupants of the others had possibly managed to force their way through the mud into the water of the swamp.

A burrow has several advantages over a house or nest on the surface. It requires far less maintenance, yet provides better insulation and more protection for both the occupant and its foodstores. The surface dwelling offers no protection against fire, and in areas which are periodically burnt, only burrowing species can survive. Most of the African grasslands are regularly swept by fire, sometimes accidentally, but usually through deliberate 'slash and burn' methods of ground clearance. In Europe and North America, firing is carried out to encourage the growth of plants useful to grouse and other game. Fire, besides threatening the surface rodent with immediate incineration, also destroys ground-cover which may take months to replace. The removal of cover is more dangerous than the actual fire since smoke and flames attract predatory birds from miles around. After a fire on one African plateau, the harsh-furred rats were almost totally annihilated and did not reappear until four months later. The pigmy mouse, *Mus triton*, however, a burrower, was trapped in greater numbers than before, although it is not known whether the mice were really more numerous or just hungrier for the bait. Studies made in Wisconsin, showed that the numbers of thirteen-lined ground squirrels and white-footed deermice were much higher in areas which were regularly burned than in unburned areas; they undoubtedly thrive on the increased yield of grass and rely on their burrows to escape from flames and predators. On the other hand, few red-backed voles, *Clethrionomys gapperi*, are found in burned areas, probably because their nests are on the surface or in bushes, and unprotected.

A rodent's basic accommodation needs can be met by a simple tunnel, with perhaps a chamber where the animal can rest or rear its young. Many species increase insulation by building a nest inside, but others such as the African spring hare, manage without one. In regions where plenty of food is available for at least part of the year, a single burrow may serve the lifetime needs of its tenant, simply being extended and modified as the need arises, sufficient food being stored in the nest or storage chamber to last through any season of scarcity. In steppe and desert areas, where home ranges tend to be large, a forager may have to travel some distance from the safety of its home burrow. In their exposed feeding grounds, species such as the thirteen-lined ground squirrel,[128] mountain marmot and some Asiatic jerboas dig short, shallow temporary burrows which serve as air-raid shelters against attack by birds of prey (Fig 33). Since neither the temporary refuges nor the breeding burrows of these species are deep enough to give protection from severe frosts, often a special winter burrow is made. The depth probably depends upon the degree of frost experienced in the area. In Nebraska, where the average depth of frost is 6cm, the hibernating nest of the thirteen-lined ground squirrel was found to be between 5 and 7cm deep, but in Manitoba, where frost penetrates 152cm, the nests were more than 2m deep.[128] The American grasshopper mouse may dig four different types of burrow: a shallow U-shaped nesting burrow, a short refuge burrow, a cache burrow (which is less than 10cm long and used for storing seeds), and finally an even shorter latrine burrow.

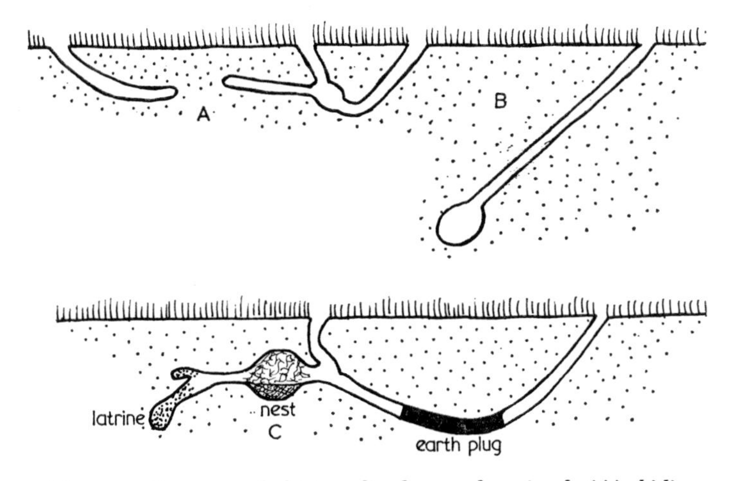

Fig 33 *Burrows of thirteen-lined ground squirrel:* (A) *hiding burrow;* (B) *hibernating burrow;* (C) *nesting burrow*

In many areas, any burrow is in some danger from floods, and the widespread practice of plugging the entrances with soil must reduce this hazard. In North Dakota, a colony of Richardson's ground squirrels was under nearly 0·5m of water for eight days, but a week after the flood subsided, the squirrels emerged from hibernation apparently unaffected by the experience.[120] The bobak marmot and black-tailed prairie dog throw out excavated soil round the burrow entrance to form a cone-shaped mound which acts as an effective dam during minor floods. Mounds can be made only when the soil is moist, and immediately after rain, the inhabitants of a dog-town busy themselves with mound-building, carrying or pushing the soil into place, then ramming it in with their noses. Burrows of the white-tailed prairie dog, which are usually on mountain slopes, are never provided with such well-maintained mounds as those of the plains-dwelling species. The burrow of the Amur suslik has an 'emergency chamber' just above the nesting cavity. If water does enter the burrow, the occupant may take refuge in this chamber; according to Ognev, the emergency chamber also helps the suslik survive extermination campaigns which involve poison gas, since the heavy gas often fails to rise up the vertical passage.

Within its burrow, a rodent is secure from most predators excepting snakes. At least one investigation of a gerbil's burrow has been rapidly terminated when two sun-snakes and a cobra, shot out in rapid succession past the mammalogist's spade. As many snakes hunt by scent, and some vipers have heat sensors which detect warm-blooded prey at a distance, an earth plug provides adequate defence against these intruders. The burrows of the African pouched rats (*Beamys* and *Saccostomus*) each have a single vertical entrance shaft which is invariably blocked at the bottom by an earth plug (Fig 34).

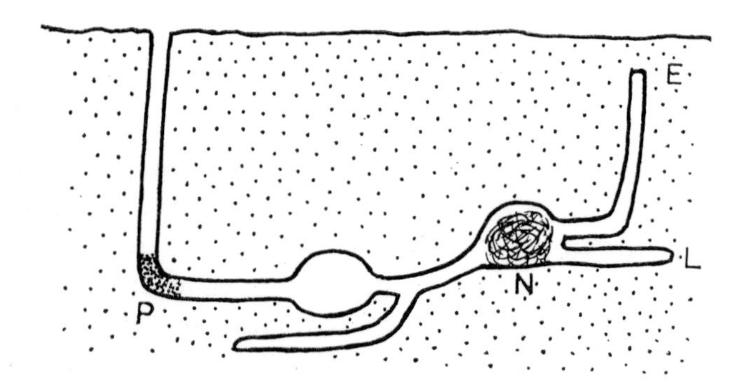

Fig 34 *Burrow of an African pouched rat*, Beamys major: (E) *emergency exit;* (L) *latrine;* (N) *nest;* (P) *earth plug*

Burrows of the African fat mouse have several exits, but they can only be located with difficulty since they are filled with soil flush to the surface. When a fat mouse is alarmed by the excavator's spade, it often flees into a side tunnel where it seals itself off before escaping through a bolt hole; similar behaviour has been reported in the American pocket mouse, *Perognathus hispidus*.[45] The tunnels of some pocket mice descend to a depth of 193cm, and those of the fat mouse to 107cm; possibly owing to the sandy soil, depth is often attained by a series of hairpin loops rather than by vertical shafts. The Australian field mouse (*Gyomys*) surrounds the open burrow entrance with small twigs and shredded vegetation; this hides the hole yet allows the occupant to get in and out with minimum inconvenience. According to Soviet workers, the burrow of the lesser jerboa, *Alactagulus pumilio*, has all the security devices of a front-line observation post. The main entrance is plugged with soil to be quite imperceptible, and the occupant is said to rest near another hole which serves as an air vent and is screened by lumps of earth; a third type of exit is blocked by a disc of soil about 2cm across and known locally as the 'belly-button'. The belly-button acts like the switch on an electric bell, for it collapses at the slightest touch and an instant later the jerboa leaps from the main exit or air vent, presumably being warned by the inrush of fresh air. The burrows of several species incorporate what appear to be emergency exits, terminating just a few centimetres from the surface.

Just as many birds have come to associate fires with food, it is a fair assumption that some are capable of recognising holes and spoil heaps as worthwhile hunting-grounds. Natural selection, especially among rodents living in open areas, has resulted in several species going to great lengths to conceal the whereabouts of their homes. The thirteen-lined ground squirrel, although it makes a plug some way down the entrance tunnel, always keeps the surrounding area free of loose soil by scattering any soil well away from the entrance, and sometimes tamping it down with its forehead.[99] The removal of spoil is a laborious process, and in several species a method of burrow construction has evolved which does not result in a tell-tale spoil heap outside the front door. After the initial descending tunnel is made, with perhaps a nesting chamber at the bottom, another tunnel is driven upwards to reach the surface some distance from the original entrance. Excavated soil is used to fill the first tunnel so the surroundings of the new hole are quite undisturbed. Eventually the old spoil heap may become overgrown with vegetation so nothing is left to draw the attention of an aerial predator. This method is used by some susliks and jerboas of the Asian steppes, the African gerbil mouse, *Malacothrix typicus*, and probably the African pouched rats (*Beamys* and *Saccostomus*). The prairie dog sometimes used this method of excavation, beginning at the dome-shaped entrance, then digging upwards to form a crater-shaped mound. In this case, the construction technique cannot be concerned with concealment, for dog-towns are visible from great distances, and the wide burrows are often used by burrowing owls and black-footed ferrets, which are both major predators.

As with the houses of wood rats, permanent burrows present hygiene problems. Most species incorporate a special latrine chamber, digging a new one when required. During early summer, the bobak marmot defaecates in a small cavity on the entrance mound, filling it with earth when full; later in the year a blind passage in the burrow serves as latrine, and at the approach of winter, its contents are used to seal the entrance. Apparently this plug sets so hard that it can be only broken by a pick, and the animal has to dig a new exit when it emerges in the spring. Nests rapidly acquire a large population of mites and various insect parasites; those of the large African pouched rats (*Cricetomys* and *Beamys*) having the unique distinction of accommodating a semi-parasitic genus of earwig (*Hemimerus*). It is probably for this reason that many burrow systems contain several nests. Marmots renew the nest lining each summer unless the pest infestation is too bad, in which case the nest chamber is sealed off and a new one made. The sanitary arrangements of some species are less commendable, for faeces and unwanted food are simply deposited in the nesting chamber. Captive steppe lemmings are notorious in this respect, for if undisturbed, the nest soon becomes a fermenting heap of soiled grass on a platform of mouldering faeces. The African mole rats (*Tachyoryctes*) use the nest chamber for sleeping, foodstores and latrine.[79] In some circumstances, this lack of hygiene, far from being detrimental to health, may have some survival value, since the heat generated by decomposition keeps the nest warm and humid.

Some species are not averse to sharing the burrow with the remains of a deceased compatriot. I have found a female fat mouse and her young sharing the nest with a badly decomposed male. The two species of prairie dogs appear to differ in their behaviour towards such unsavoury co-tenants. The introduction of dead animals into burrows of the black-tailed species, resulted in the burrow openings rapidly being sealed unless the carcass could be walled off in a side passage.[148] The white-tailed prairie dog was found to be unperturbed by the presence of a corpse in its living quarters. Early writers have suggested that prairie dogs entomb any rattlesnakes they find in their burrows. This story was exploded by William Longhurst of Colorado who tethered four rattlesnakes by their tails inside occupied burrows for four days. The dogs, perhaps displaying more discretion than the researcher, remained inside for as long as a snake blocked the exit, and there was never any attempt to entomb the intruder. After the snakes were removed, the burrows were deserted for several days.

12

RODENTS AND MAN

The evolution of man from hunter to cultivator must have had an effect on the rodent world nearly as profound and far-reaching as it had on his own. In many areas, some species—especially seed-eaters—no longer had to search for wild plants, since cultivated crops ensured a regular and plentiful food supply. As the desert was irrigated and brought under the plough, and forest replaced by arable land, enormous new areas were opened up to those relatively unspecialised rodents able to exploit them.

The association of rodents with cereal crops has been well-known since ancient times. About the sixth century BC, a coin of the Greek colony of Metapontum figured a harvest mouse climbing an ear of barley (Fig 35). At the other side of the world, in Peru, Inca and pre-Inca pottery have incised representations of rats and mice gnawing ears of corn. Rodents have been responsible for the destruction of incredible quantities of grain. In 1932, a Russian authority estimated that a single suslik damaged 4kg of cereals each year; the total population of 600,000,000 could account for 2,538,640 tonnes of grain—an amount equivalent to about one-sixth of the total harvest of the United Kingdom in 1971. Similar figures could be prepared for other species, but it has already been pointed out how a rodent's diet varies according to season and locality, and how difficult it is to obtain reliable estimates of numbers.

The rapid entry of rodents into cultivated land has been especially noticeable during the present century, for virgin lands can be opened up so quickly, and governments and farmers alike are conscious of the need to exact the maximum possible yield. One of the most injurious rodents in Manchuria is the giant rat-headed hamster, *Cricetulus triton*; it penetrated the country after the Chinese Eastern Railway was constructed in 1898, and the forests cleared for agriculture. According to Loukashin, this animal is as formidable as its name, being untameable, and attacking and killing other small rodents. Near Harbin, the Manchurian striped hamster disappeared after the rat-headed species moved in. During the mating season, males are said to attack each other, the victor sometimes eating the vanquished. Notwithstanding its destructive habits, this hamster was probably regarded with some favour by the poorer Chinese peasants who sometimes eked out a living by digging up the rodent's grain stores. Nowadays, not even this 'Robin Hood' quality has any place in a People's Republic.

Fig 35 *Ancient Greek coin, showing a mouse and an ear of barley*

In at least the developed countries of the world, the era of the cornfield rodent is drawing to a close. Until recently, it found easy gleanings in the stubble fields, and comfortable winter quarters in the rick yards. During an eight-month study carried out on a farm in southern England in 1959–60, 563 harvest mice were taken from 36 poison-baited ricks. Today, it would be difficult to find a corn-rick anywhere in England. The countryside is now dominated by the steel silo, which, together with the combine harvester and the practice of burning straw and stubble, make the cereal farm an extremely uncomfortable habitat for any rodent. Of course, cereals are not the only crops to be attacked by rodents. Almost any kind of edible material, root, stem or fruit, is liable to be damaged. Cash crops which are often subjected to particularly heavy depredations by rodents include groundnuts and sugar cane. In some regions, rats and gerbils destroy acres of cotton in their attempts to extract the oily seeds from the pods.

Whereas the introduction of arable farming could not fail to work to the advantage of the grain-feeding and omnivorous species, the raising of livestock had a variable effect on grazing rodents. In some regions, such as Europe, the replacement of forests by sheep pastures allowed voles to extend their range; rodents everywhere benefited by the control measures taken against the potential predators of livestock. In regions where grassland already existed, grazing rodents were, except in South America, secondary in ecological importance to the herbivores, and it is unlikely that the change from bison and zebra to cattle and sheep affected them significantly.

From the farmer's point of view, the effect on grassland of voles, lemmings and grass mice is generally negligible when compared with the damage caused by rabbits. Nevertheless peak populations of small rodents can, of course, seriously reduce grazing for livestock. Densities of over 2,470 voles per hectare sometimes occur, while in Nevada a *Microtus* density of over 29,500 per hectare has been recorded. Some of the larger, colonial species have a more noticeable effect on the environment, and consequently are regarded as pests. In the plains of North America, a campaign has been waged against prairie dogs since the time of the early settlers, and a few years ago, conservationists became concerned for the animal's survival. In Kansas alone, the area occupied by prairie dogs, fell from 1,015,625 hectares in 1903, to 23,532 hectares in 1957. The dogs certainly feed on grasses useful to cattle, and their towns are always in overgrazed areas, just as they were when they shared the range with the bison. However, as Professor Smith pointed out in his monograph on the prairie dog, the abundance of various rodents on depleted ranges is the result, instead of the cause, of overgrazing. After cattle were removed from an area occupied by a prairie dog colony in Oklahoma, the grass increased to such an extent that the dogs left, apparently unable to adapt to the altered habitat. Many ranchers now realise that instead of waging extermination campaigns, they can limit the numbers of prairie dogs by ensuring a good grass crop through proper cattle management. In the South American pampa, rodents were the dominant grazing animals until horses were introduced in 1536 and cattle in 1556. Here, the most serious pest is the viscacha which consumes large quantities of grass, and like the prairie dog, so undermines the ground that horses and cattle sometimes crash through and break their legs.

The rapid destruction of the world's natural woodlands presents a grave threat to many forest species which are too highly specialised to adapt to change. The great evergreen forest which once extended across the centre of Africa now exists as small relict stands flanking a few rivers and capping some mountains. When these finally disappear, several forms such as the scaly-tails and pouched rats (*Beamys*) will become extinct. Undoubtedly, many forest-dwellers in other parts of the world are in the same situation. Other forest species have less exacting requirements, and like seed-eating rodents presented with a cornfield, adapt with alacrity to life in a new plantation. With a few exceptions, rodents rarely cause more than minor damage to mature timber, most destruction is through gnawing the bark of newly planted seedlings and digging up seeds. In parts of North America the porcupine is achieving the status of a pest, and in Britain, the introduced grey squirrel is officially regarded as a menace second only to the rat. It causes serious damage by stripping the bark of certain trees, such as sycamore, during the summer months

—a habit which seems associated with social behaviour rather than dietary needs. In 1973, the British government, having subsidised the destruction of squirrels for many years, produced a White Paper, *The Grey Squirrel (Warfarin) Order*, which allows the use of anti-coagulent poison against squirrels, subject to certain safeguards.

In Britain, the native red squirrel is officially regarded with more affection, probably because it is comparatively rare. Towards the end of the last century, thousands of red squirrels were shot when their numbers reached plague proportions, and plantations and gardens were ravaged. In its native American hardwoods, the grey squirrel is rarely regarded in an unfavourable light. It is an important game animal, protected during the breeding season and often encouraged by the provision of nesting-boxes. Even in the USA, however, its record is not entirely without blemish. In the north-east, it has been reported to cause $1,000,000 worth of damage annually, through a propensity for gnawing the insulation from electric cables.

A few species, such as the American grasshopper mouse and the African harsh-furred rat, are unquestionably beneficial to agriculture through their destruction of insects. Others can be useful in improving soil fertility; pocket gophers and other burrowing species may destroy a certain number of root crops and newly planted trees, but they also turn over considerable amounts of soil, and improve drainage and aeration. Following his observations on northern pocket gophers, *Thomomys talpoides*, in Utah, one investigator calculated that thirty animals would move 38·6 tonnes of soil annually.[127] Another worker estimated that pocket gophers raised 8,129 tonnes of earth in the Yosemite Park, California, each year. In another part of California, it was reported that kangaroo rats, through scattering the soil around their burrows, caused a five-fold increase in the amount of sheep forage. It is difficult to calculate the effects of rodent burrows on soil drainage, but a single burrow of an African mole rat was found to take 5,232·56 litres of water poured at 20 litres per minute.

While many rodents are regarded as pests by more prosperous communities, they are highly valued as food in underdeveloped countries where malnutrition is rife. The situation in many parts of Africa has scarcely changed since 1855, when Livingstone wrote, 'Moles and mice constitute important articles of diet among them; and traps may be seen fringing the paths for miles together at intervals of 10 or 15 yards.' In some regions, rodents are the only source of animal protein. The giant pouched rat is highly sought after throughout its range, sometimes being smoked and used as storage food. During the 1930s an attempt was made in the Belgian Congo to rear this species on a commercial scale. At one time, European palates were partial to rodent meat. About 100 BC, the fat dormouse, *Glis glis*, was a luxury dish for wealthy Romans. The animals were reared in special enclosures called *gliaria* which were planted with oak and beech. Before fulfilling their ultimate destiny, the plumpest specimens were transferred to small earthenware vessels, or *dolia*, in which they were fattened on acorns and chestnuts—a method of 'sweatbox' treatment which was to be adopted for other livestock some 2,000 years later. Nowadays, only the larger rodents such as squirrels, coypu and muskrat are eaten to any extent in Europe. In North America, muskrat carcasses are often sold for eating under the more acceptable name of 'marsh rabbit'.

Few Europeans or Americans would entertain the idea of eating guinea pig, yet this was the first rodent to be domesticated, and was one of the Inca's principal sources of meat. Excavations of a Peruvian cave used about 7,000 years ago by Palaeolithic man, revealed numerous bones of guinea pig, but it is uncertain whether they belonged to domesticated or wild specimens. About twenty species of guinea pig are distributed over the northern half of South America, but the numerous domestic strains all belong to a single species, *Cavia porcellus*. The domesticated guinea pig was introduced into Europe about 1580, probably by returning colonists or seamen who appreciated its virtues, either as a pet or as a change from salt pork. In seventeenth-century Germany, guinea pigs were sometimes served at table. The origin of the English name is obscure; it could be a corruption of Guiana pig, or perhaps the animals reached England via slaving vessels operating between South America and the African Guinea Coast. One of the first English slavers was Sir John Hawkins, whose son, Sir Richard, was the first Englishman to describe another South American rodent which was destined to become of greater economic importance than the guinea pig. In *The Observations of Sir Richard Hawkins in his voyage into the South Sea in the year 1593*, he recorded, 'Amongst others, they have little beastes like unto a squirrell, but that hee is gray; his skinne is the most delicate, soft, and curious furre that I have seene, and of much estimation (as is of reason) in the Peru; few of them come into Spaine, because difficult to come by; for that the princes and nobles laie waite for them. They call this beast chinchilla, and of them they have great abundance.' Chinchilla fur is not only one of the finest and softest in the world, but is very light in weight and has a beautiful bluish-grey colour. It became extremely popular during the nineteenth century. The London auction rooms of the Hudson's Bay Co began dealing in the skins in 1842, sometimes selling over 200,000 a year. At the turn of the century, wild chinchillas became so rare that trapping became uneconomic, and South American countries attempted to conserve the survivors and establish fur farms. The introduction of chinchilla farming to North America and Europe is due to an American mining engineer, M. F. Chapman, who, in 1922 obtained a licence to trap and export living specimens. By bringing his captives down in easy stages from the high Andes, he succeeded in acclimatising them to lower altitudes. The present-day chinchilla industry was built up from the descendants of Chapman's three females and eight males. Over 150 pelts are needed to make a full-length coat, each pelt selling today for about £20.

About the time that chinchilla fur became popular, Argentina realised the economic potential of the coypu which was so common in its rivers and marshes. In the coypu, or nutria (the Spanish word for otter), as in most rodents, it is the soft underfur which is used by the fur trade. The animals were so plentiful that 500,000 skins were exported each year, reaching over 1,000,000 in 1918. Thereafter the numbers declined, and as the governments of Argentina and Uruguay took steps to prevent the extinction of a major source of revenue, overseas speculators became interested in rearing the animals. A few attempts had previously been made in France in 1882, and in California in 1899, but the real boom in nutria farming took place in the 1920s and 1930s when they were introduced into most European countries, Russia and Canada. Thousands of animals were imported into the USA during the 1950s, when promoters agreed to buy offspring at $380 a pair provided the demand continued. When the market collapsed, many disappointed farmers liberated their stock into the nearest stream. The coypu is now well established in the waterways of the Pacific States, Russia and Europe. Although the species is in danger of becoming extinct in many parts of its native range, it has thrived in most places of introduction. In some countries, the proliferation of escaped and liberated animals was welcomed. In France, for instance, it was given protection as a game animal in 1939. The coypu is regarded with less affection by the British, owing to its damage to sugar-beet. For many years the numbers of British coypu remained small and confined to a single river system, but a succession of mild winters led to an increase in numbers and extensions of range. In 1972, an estimated 11,000–12,000 coypu in East Anglia led to an intensification of the government-backed control campaign. No nutria farms exist in Britain today, but the animals are extensively reared in Czechoslovakia, Poland and Russia.

Unlike nutria, the fur of the muskrat, or musquash, continues to find a ready market in America and western Europe. The major firm of London dealers, Hudson's Bay and Annings, handles 500,000 each year. The species was introduced into Europe by enthusiastic fur farmers, but nowadays most commercial furs are obtained from preserves in Canada, and it is doubtful whether any European farmer could compete with the present 1974 price of £1 per undressed pelt. In Europe, the progeny of farm escapees have become far greater pests than coypu, by damaging river banks. The muskrat enjoyed a very short stay in Britain. It was introduced into Ireland in 1927, and England and Scotland in 1929; by 1937 it had been exterminated from all areas. The keeping of muskrats in Britain is now prohibited. The Belgian government has been less fortunate. In spite of earlier poisoning and trapping campaigns, forty trappers caught 133,413 specimens in 1971, but numbers now appear to be declining.

The most important fur-bearing rodent is the beaver. In North America, the economy of the forest Indian was centred on this animal; its skin was used for clothing, fat as protection against frostbite, incisor teeth as edge tools, and the castoreum as a general cure-all and additive to tobacco. To European cultures, beaver fur represented only a source of profit, and no consideration was given to the animal's survival. The last authentic record of beavers in the British Isles was made by a traveller in Wales, who, in 1188, stated that the Teivi in Cardiganshire was the only river in which they occurred. Within a few centuries, beavers became rare throughout Europe as a result of the demand for their furs. Beaver hats were especially popular. As early as 1386, Chaucer referred to a 'Flaundrish Bever hat', and the fashion continued for several centuries. Luckily for the cranial comforts of British and French gentlefolk, the decline of the European beaver did not put an end to the hat industry. It was saved by the Cabots and Cartier, who, in their search for the 'north-west passage', discovered a new source of furs on the other side of the Atlantic. Both Britain and France were anxious to obtain the revenue from a new fur trade, and established trading settlements and townships. In 1670, 'the governor and company of adventurers of England trading to Hudson's Bay' obtained a charter from Charles II. Then the exploitation of the American beavers began in earnest (leading eventually to the creation of the Dominion of Canada). In 1743, when the French still enjoyed the monopoly of trade, 127,000 beaver skins were imported into La Rochelle in France. After the market was transferred to London, annual sales rarely fell below 100,000, and between 1860 and 1880, 3,500,000 skins were sold. In the early years of fur-trading, beaver pelts were used as currency, a certain number being traded in exchange for a blanket or gun. The Dutch traders of New Amsterdam (New York) were less scrupulous, taking skins from Indians in return for wampum, strings of shell beads. Judging by the cost of skins in Europe, enormous profits must have been made; Samuel Pepys recorded in his diary for 27 June 1661, that he paid £4 5s for a beaver hat. The beaver was saved from annihilation possibly when silk hats were introduced during the last century. It is now protected throughout its range, and is reared extensively on Canadian preserves. Surprisingly, the trade in skins is as vigorous as it was 200 years ago. A yearly average of 120,000 skins is offered by the major London dealers, at prices between £8 and £20 for a large, undressed pelt. The only other rodent of interest to the London market is the Canadian squirrel. In Asia, marmots and susliks have long been a valuable source of furs and meat, the fat of the bobak marmot also being used for oiling shoe leather and for veterinary purposes.

In nineteenth-century London, the brown rat played a part in the sporting life of the city. As an attraction for their clientèle, several public houses arranged rat matches at regular intervals. The purpose of the match was to find which of the customers' dogs could kill the most rats in the shortest possible time. The usual procedure consisted of releasing fifty rats into a small arena, then throwing in a dog and timing it with a stop-watch. Naturally, the main interest was in the wagers which were laid on the favourite dogs. One of the most celebrated London ratters, 'Billy', managed to kill 500 rats in five and a half minutes. Probably it was from this sport that the term 'rat-race' originated, for the rats which raced round the arena and dodged the first dog, would invariably perish in a subsequent match. According to Mayhew's contemporary account, the proprietor of one of the largest sporting houses in London, purchased 500 rats each week, and had twenty families of rat-catchers depending on him for a living. Farmers welcomed the sport since they no longer had to pay for rats caught on their land, rat-catchers selling them to a sporting landlord at 3d per head. Warehousemen were doubly fortunate, receiving payment from both the warehouse owner and the landlord. Before the flushing of the sewers became general, sewermen augmented their wages by rat-catching, but their contributions were not over-popular with the sporting fraternity. According to experts of the time, 'sewer rats are dreadful for giving dogs canker of the mouth', and 'the smell that arose from them was like that from a hot drain'.

Numerous rodent species act as reservoirs for organisms responsible for serious diseases. Of all the ills which afflict mankind, plague is one of the most terrifying, producing a mortality of up to 95 per cent among those infected. The first recorded great plague erupted in the eastern part of the Roman Empire in 540; the second, in the fourteenth century, affected the whole of Europe, parts of Asia, Africa, and even Greenland. This epidemic, given the dramatic, if inaccurate, name of 'Black Death' by a writer in 1823, reached England in 1348 and is believed to have killed about a twentieth of the population.[147] In spite of subsequent epidemics, including that which struck London in 1665 and reduced the capital's population to half its normal size, the bacillus responsible was not discovered until 1894, and the parts played by flea and rat were not fully elucidated until 1914. If a flea feeds on a plague-infected rat, its stomach becomes blocked by a blood clot containing a pure culture of bacillus, and it dies within about three days. If, in the meantime, the infected flea attempts to feed on another animal, bacteria are regurgitated and infect the new host. Most epidemics of bubonic plague involve the domestic rats, *Rattus rattus* and *R. norvegicus*, whose fleas are capable of penetrating the human skin with their mouthparts. In areas where plague is endemic, rats may acquire considerable immunity. If the rats succumb to the disease, fleas leave their bodies and attack humans; as early as 1603, it was noticed that rats and mice left their holes at the time of plague. In medieval Europe, rats and fleas found ideal living conditions in the thatched, wattle and daub houses, just as they do today in the primitive dwellings of underdeveloped countries. In temperate climates, plague outbreaks generally subside in winter when the fleas hibernate in the rats' nests.

Although plague is usually transmitted to man through domestic rats, the disease exists in a smouldering state over enormous areas of steppe and semi-desert country where several species of wild rodents serve as reservoirs. Wild-rodent plague rarely causes much human mortality, but it prevents total eradication of the disease; wild rodents may, at any time, infect domestic rats, which in turn cause an outbreak in the human population. Ground squirrels are an important reservoir in California; in New Mexico, prairie dogs and marmots are infected, and in South America, guinea pigs may be involved. In South Africa, most human infections are contracted in farm huts infested with domestic rodents which have acquired the disease from gerbils. Out of 900 outbreaks reported between 1919 and 1943, all but one were in rural areas, the majority on farms.[37] At the time of an outbreak, it is typical to find that the wild rodent colonies have vanished, and the domestic species greatly reduced. Apparently plague can sweep through a rodent population without causing a single infection in man. In Siberia and Mongolia, trappers and furriers are especially at risk, for both susliks and marmots harbour pneumonic plague. At low temperatures, certain fleas of these wild rodents are able to survive without feeding for up to 369 days; consequently the disease can be spread to distant areas through fleas left inside the pelts.

In warm countries, various species of rodents act as reservoirs for the protozoan parasites known as Leishmanias, which are responsible for oriental sore in Asia, espundia and Chaga's disease in South America. In central Asia, oriental sore is common in isolated settlements on the desert fringes. Here, the disease is endemic among the wild gerbils and is transmitted by infected sandflies living in the burrows. In one settlement in Turkmenistan, the poisoning of 500,000 gerbil burrows succeeded in reducing the incidence of the disease from 70 per cent to almost zero within a year.[69] Rodents are also reservoirs for several forms of typhus—scrub typhus of the Far East, endemic typhus which has a worldwide distribution, and Rocky Mountain fever, which, in spite of its name, is found over most of North America. Fortunately these diseases are less dangerous than epidemic or louse-borne typhus, but Rocky Mountain fever can produce a high mortality unless treated. This disease is transmitted by ticks. Pocket gophers and voles have been incriminated as reservoirs, but ground squirrels and other mammals also could be involved.

A few virus diseases may be transmitted directly through the bite of an infected rodent. Rat-bite fever is common in Japan, while in Thailand about 4 per cent of the two species of domestic rats have been found infected with rabies. Although the rodent enthusiast has a very remote chance of contracting any of these diseases, *Leptospirosis* is a virus disease which presents a real hazard. The virus has been recorded in several species of rodent, and in other mammals. It seems to cause no harm to the host and is excreted in the urine. Contaminated water may be a source of infection among sewer workers and bathers, infections due to rat bites have also been reported. Anyone handling rodents, or cleaning their cages, should wash his hands afterwards and be especially careful to prevent contamination of cuts.

The sum total of human misery due to rodents and to rodent-borne disease is incalculable. Nevertheless, besides playing an essential role in the ecology of numerous plant and animal communities, even the most maligned species are now contributing to man's welfare. In May 1973, it was reported in the British House of Commons that over 5,500,000 experiments on living animals took place in the country each year, and that 95 per cent of the animals used were rats, mice and guinea pigs. It is sometimes forgotten that the phenomenal progress of medical research would have been impossible without the use of rodents as experimental animals.

APPENDIX

CLASSIFICATION: THE FAMILIES OF RODENTS

The classification of the rodents has been a subject of dissension among taxonomists for many years. The true affinities of some members of this diverse order will probably never be known. There is general agreement on classification to family level, although there is no clear dividing line between the Muridae and Cricetidae. The outline given below is intended only as a means of placing some of the species mentioned in the text. Super-family groups have been omitted. For more detailed information, reference should be made to the following: Ellerman, J. R., *The families and genera of living rodents* (1940–9), and Simpson, G. C., *The principles of classification and a classification of mammals* (1945). Desmond Morris, in *The mammals* (1965) has produced an excellent summary based on the works of several taxonomic authorities.

Order RODENTIA

Suborder SCIUROMORPHA (Squirrel-like rodents)

Family 1 APLODONTIDAE

One species. Mountain beaver (*Aplodontia rufa*) (N. America)

Family 2 SCIURIDAE

Subfamily 1 SCIURINAE (N. and S. America, Europe, Asia, Africa)

Tree squirrels, ground squirrels, marmots, prairie dogs, chipmunks

Subfamily 2 PETAURISTINAE (Asia, N. and Central America)

Flying squirrels

Family 3 GEOMYIDAE (N. and Central America)

Pocket gophers

Family 4 HETEROMYIDAE (N., S. and Central America)

Pocket mice (*Perognathus*); kangaroo mice (*Microdipodops*); kangaroo rats (*Dipodomys*); spiny pocket mice (*Liomys* and *Heteromys*)

Family 5 CASTORIDAE (N. America, Europe, Asia)

Beaver (*Castor fiber*)

Family 6 ANOMALURIDAE (Africa)

African scaly-tailed squirrels

Family 7 PEDETIDAE (Africa)

Spring hare (*Pedetes capensis*)

Suborder MYOMORPHA (Mouse-like rodents)

Family 8 CRICETIDAE

Subfamily 1 CRICETINAE (N., Central and S. America)

About 56 genera of New-World mice. Includes wood rats (*Neotoma*); grasshopper mice (*Onychomys*); fish-eating rats (*Ichthyomys*); hamsters; mole rats or zokors (*Myospalax*)

Subfamily 2 NESOMYINAE (Madagascar)

Malagasy rats

Subfamily 3 LOPHIOMYINAE (Africa)

Maned rat (*Lophiomys imhausi*)

Subfamily 4 MICROTINAE (N. and Central America, Europe, Asia, Africa, Arctic)

Lemmings; voles; muskrats; mole lemmings

Subfamily 5 GERBILLINAE (Asia, Africa)

Gerbils

Subfamily 6 DENDROMURINAE (Africa)

Includes climbing mice (*Dendromus*); fat mice (*Steatomys*)

Subfamily 7 CRICETOMYINAE (Africa)

Pouched rats (*Cricetomys*, *Beamys* and *Saccostomus*)

Subfamily 8 PETROMYSCINAE (Africa)

Rock mice (*Petromyscus*); swamp mouse (*Delanymys brooksi*)

Subfamily 9 OTOMYINAE (Africa)

Swamp rats (*Otomys*); karroo rats (*Parotomys*)

Family 9 SPALACIDAE (N. Africa, Europe, Asia)

Mole rat (*Spalax*) (possibly a single species)

Family 10 RHIZOMYIDAE (E. Africa, S. Asia)

E. African mole rats (*Tachyoryctes*); bamboo rats (*Rhizomys*); lesser bamboo rat (*Cannomys badius*)

Family 11 MURIDAE

Subfamily 1 MURINAE (Europe, Asia, Africa, Australian region)

About 80 genera of Old-World rats and mice. Includes domestic rats (*Rattus*); house mice (*Mus*); African water rats (*Dasymys*); harsh-furred rats (*Lophuromys*); spiny mice (*Acomys*)

Subfamily 2 PHLOEOMYINAE (Asia)

Confined to the Far East. Includes prehensile-tailed rats (*Pogonomys*)

Subfamily 3 RHYNCHOMYINAE

Philippines shrew rat (*Rhynchomys soricoides*)

Subfamily 4 HYDROMYINAE (Australian region)

Water rats of Australia, New Guinea and the Philippines. Includes *Xeromys*, *Hydromys*, *Mayermys*, *Baiyankamys*

Family 12 GLIRIDAE

Subfamily 1 GLIRINAE (Europe, Asia)

Dormice

Subfamily 2 GRAPHIURINAE (Africa)

Dormice (*Graphiurus*)

Family 13 PLATACANTHOMYIDAE (S. and S.E. Asia)

Two species of spiny dormice of Southern India, China and Indo-China

Family 14 SELEVINIIDAE (Central Asia)

Desert dormouse (*Selevinia betpakdalaensis*)

Family 15 ZAPODIDAE

Subfamily 1 SICISTINAE (N. Europe, Asia)

Birch mice (*Sicista*)

Subfamily 2 ZAPODINAE (N. America, E. Asia)

American jumping mice (*Zapus* and *Napaeozapus*); Chinese jumping mouse (*Eozapus setchuanus*)

Family 16 DIPODIDAE (Asia, N. Africa)

Jerboas

Suborder HYSTRICOMORPHA (Porcupine-like rodents)

Family 17 HYSTRICIDAE (Old-World porcupines)

Subfamily 1 HYSTRICINAE (Africa, Asia, Indonesia)

Large porcupines (*Hystrix*); Indonesian porcupines (*Thecurus*)

Subfamily 2 ATHERURINAE (Africa, S. Asia, Indonesia)

Brush-tailed porcupines (*Atherurus*); long-tailed porcupine (*Trichys lipura*)

Family 18 ERETHIZONTIDAE (New-World porcupines)

Subfamily 1 ERETHIZONTINAE (N., Central and S. America)

North American porcupine (*Erethizon dorsatum*); tree porcupines (*Coendou*); Upper Amazon porcupine (*Echinoprocta rufescens*)

Subfamily 2 CHAETOMYINAE (S. America)

Thin-spined porcupine (*Chaetomys subspinosus*)

Family 19 CAVIIDAE

Subfamily 1 CAVIINAE (S. America)

Includes guinea pigs (*Cavia*)

Subfamily 2 DOLICHOTINAE (S. America)

Mara (*Dolichotis patagonum*); salt-desert cavy (*Pediolagus salinicola*)

Family 20 HYDROCHOERIDAE (S. America)

Capybara (*Hydrochoerus hydrochaeris*)

Family 21 DINOMYIDAE (S. America)

False paca (*Dinomys branickii*)

Family 22 DASYPROCTIDAE

Subfamily 1 CUNICULINAE (Central and S. America)

Paca (*Cuniculus paca*); mountain paca (*Stictomys taczanowskii*)

Subfamily 2 DASYPROCTINAE (Central and S. America)

Agoutis (*Dasyprocta*); acouchis (*Myoprocta*)

Family 23 CHINCHILLIDAE (S. America)

Chinchillas; mountain viscachas (*Lagidium*); viscacha (*Lagostomus maximus*)

Family 24 CAPROMYIDAE (W. Indies, Cuba, Central and S. America)

Hutias (*Capromys*, *Geocapromys* and *Procapromys*); zagoutis (*Plagiodontia*); coypus (*Myocastor coypus*)

Family 25 OCTODONTIDAE (S. America)

Includes cururo (*Spalacopus cyanus*)

Family 26 CTENOMYIDAE (S. America)

Tucu-tucos (*Ctenomys*)

Family 27 ABROCOMIDAE (S. America)

Rat chinchillas (*Abrocoma*)

Family 28 ECHIMYIDAE (Central and S. America)

South American spiny rats. Includes *Echimys*

Family 29 DACTYLOMYIDAE (S. America)

South American climbing rats. Includes *Dactylomys*

Family 30 THRYONOMYIDAE (Africa)

Cane rats (*Thryonomys*)

Family 31 PETROMYIDAE (Africa)

Rock rat (*Petromus typicus*)

Family 32 BATHYERGIDAE (Africa)

Mole rats (*Georychus capensis*, *Cryptomys*, *Heliophobius* and *Bathyergus*); naked mole rat (*Heterocephalus glaber*)

Family 33 CTENODACTYLIDAE (Africa)

Gundis

NOTES AND REFERENCES

1 Aleksink, M. 'The function of the tail as a fat storage depot in the beaver (*Castor canadensis*).' *J Mammal*, 51 (1970), 145–8

2 Armitage, K. B. 'Vernal behaviour of the yellow-bellied marmot (*Marmota flaviventris*)', *Anim Behav*, 13 (1965), 59–68

3 Baker, R. H. 'Nutritional strategies of myomorph rodents in North American grasslands', *J Mammal*, 52 (1971), 800–5

4 Baring Gould, S. *Curious myths of the middle ages* (1875)

5 Barlow, J. C. 'Observations on the biology of rodents in Uruguay', *Life Sci Contr R Ont Mus*, 75 (1969)

6 Barnett, S. A. 'Social behaviour in wild rats', *Proc Zool Soc Lond*, 130 (1958), 107–52

7 Barnett, S. A. *A study in behaviour* (1963)

8 Barnett, S. A. and Widdowson, E. M. 'Organ-weights and body-composition in mice bred for many generations at − 3° C', *Proc R Soc*, 162 (1965), 502–16

9 Bartholomew, G. A. and Hudson, J. W. 'Desert ground squirrels', *Scientific American* (Nov 1961), 107–16

10 Batchelder, C. F. 'Notes on the Canadian porcupine', *J Mammal*, 29 (1948), 260

11 Batzi, G. O. and Pitelka, F. A. 'Condition and diet of cycling populations of the California vole, *Microtus californicus*', *J Mammal*, 52 (1971), 141–63

12 Beck, A. M. and Vogl, R. J. 'Effects of spring burning on rodent populations in a brush prairie savanna', *J Mammal*, 53 (1972), 336–46

13 Bintz, G. L. 'Tissue catabolism of laboratory rats and *Spermophilus lateralis* during acute negative water balance', *J Mammal*, 50 (1969), 355–6

14 Blake, B. H. 'The annual cycle and fat storage in two populations of golden-mantled ground squirrels', *J Mammal*, 53 (1972), 157–67

15 Blus, L. J. 'Relationship between litter size and latitude in the golden mouse', *J Mammal*, 47 (1966), 546–7
16 Bowers, J. M. and Alexander, B. K. 'Mice: Individual recognition by olfactory cues', *Science*, 158 (1967), 1208–10
17 Bradley, W. G. 'Food habits of the antelope ground squirrel in southern Nevada', *J Mammal*, 49 (1968), 17–21
18 Bradley, W. G. and Mauer, R. A. 'Reproduction and food habits of Merriam's kangaroo rat, *Dipodomys merriami*', *J Mammal*, 52 (1971), 500–7
19 Brander, R. B. and Books, D. J. 'Return of the fisher', *Natural History*, 82 (1973), 52–7
20 Bridgewater, D. D. 'Predation of *Citellus tridecemlineatus* on other vertebrates', *J Mammal*, 47 (1966), 345–6
21 Broadbooks, H. E. 'Home ranges and territorial behavior of the yellow-pine chipmunk, *Eutamias amoenus*', *J Mammal*, 51 (1970), 310–26
22 Brooks, R. J. and Banks, E. M. 'Behavioural biology of the collared lemming', *Anim Behav Monag*, 6 (1973)
23 Brown, J. C. and Twigg, G. I. 'Some observations on grey squirrel dreys in an area of mixed woodland in Surrey', *J Zool Lond*, 144 (1965), 131–4
24 Bryan, C. P. *The papyrus ebers*
25 Cameron, G. N. and Rainey, D. G. 'Habitat utilization by *Neotoma lepida* in the Mohave Desert', *J Mammal*, 53 (1972), 251–66
26 Carleton, W. M. 'Food habits of two sympatric Colorado sciurids', *J Mammal*, 47 (1966), 91–103
27 Carr, W. J. and Caul, W. F. 'The effects of castration in rat upon the discrimination of sex odours', *Anim Behav*, 10 (1961), 20–7
28 Christian, J. J. 'Fighting, maturity, and population density in *Microtus pennsylvanicus*', *J Mammal*, 52 (1971), 556–67
29 Clough, G. C. 'Biology of the Bahamian hutia (*Geocapromys ingrahami*)', *J Mammal*, 53 (1972), 807–23
30 Clulow, F. V. and Clarke, J. R. 'Pregnancy block in *Microtus agrestis* and induced ovulator', *Nature*, 219 (1968), 511
31 Coles, R. W. 'Pharyngeal and lingual adaptation in the beaver', *J Mammal*, 51 (1970), 424–5
32 Corbet, G. B. *The terrestrial mammals of Western Europe*, 1966
33 Cowley, J. J. and Wise, D. R. 'Some effects of mouse urine on neonatal growth and reproduction', *Anim Behav*, 20 (1972), 499–506
34 Crouch, G. L. 'Clipping of woody plants by mountain beaver', *J Mammal*, 49 (1968), 151–2

35 Crowcroft, P. and Rowe, F. P. 'The growth of wild mouse colonies', *Proc Zool Soc Lond*, 129 (1957), 359–70
36 Crowcroft, P. and Rowe, F. P. 'Social organisation and behaviour in the house mouse', *Proc Zool Soc Lond*, 140 (1963), 517–31
37 Davis, D. H. S. 'Sylvatic plague in South Africa: History of plague in man 1919–1943', *Ann Trop Med Parasit*, 42 (1948), 207–17
38 De Beer, B. 'Rodent moles of Randfontein', *Afr Wild Life*, 19 (1965), 243–9, 331–4
39 De Kock, L. L. and Robinson, A. E. 'Observations on a lemming movement in Jamtland, Sweden, in autumn 1963', *J Mammal*, 47 (1966), 490–9
40 Delany, M. J. 'The biology of small rodents in Mayanja Forest, Uganda', *J Zool Lond*, 165 (1971), 85–129
41 Dixon, A. K. and Mackintosh, J. H. 'Effects of female urine upon the social behaviour of adult male mice', *Anim Behav*, 19 (1971), 138–40
42 Doty, R. L. and Kart, R. 'A comparative and developmental analysis of the midventral sebaceous gland in 18 taxa of *Peromyscus*, with an examination of gonadal steroid influences in *Peromyscus maniculatus bairdii*', *J Mammal*, 53 (1972), 83–99
43 Drummond, D. C. 'Variation in rodent populations in response to control measures', *Zool Soc Symp*, 26 (1970), 351–67
44 Eibl-Eibesfeldt, I. 'Das Verhalten der Nagetiere', *Handb Zool*, Berlin, Bd 8, Lfg 12, Teil 10 (1958), 1–88
45 Eisenberg, J. F. 'The behavior of heteromyid rodents', *Univ Calif Publi Zool*, 69 (1963), 1–114
46 Eisenberg, J. F. and Isaac, D. E. 'Reproduction of heteromyid rodents in captivity', *J Mammal*, 44 (1963), 61–7
47 Eloff, G. 'Adaptation in rodent moles and insectivorous moles and the theory of convergence', *Nature*, 168 (1951), 1001
48 Elton, C. *Voles, mice and lemmings: Problems in population dynamics* (1942)
49 Ewer, R. F. 'Food burying in the African ground squirrel, *Xerus erythropus*', *Z Tierpsychol*, 22 (1965), 321–7
50 Ewer, R. F. 'Behaviour of the African giant rat (*Cricetomys gambianus* Waterhouse)', *Z Tierpsychol*, 24 (1967), 6–79
51 Ewer, R. F. 'Biology and behaviour of a free living population of black rats (*Rattus rattus*)', *Anim Behav Monag*, 4 (1971)
52 Fall, M. W., Medina, A. B. and Jackson, W. B. 'Feeding patterns of *Rattus rattus* and *Rattus exulans* on Eniwetok Atoll, Marshall Islands', *J Mammal*, 52 (1971), 69–76

53 Fertig, D. S. and Edmonds, V. W. 'The physiology of the house mouse', *Scientific American* (1969), 103–10
54 Fisher, J. 'A captive mountain beaver', *J Mammal*, 46 (1965), 707–8
55 Fleming, T. 'Notes on the rodent fauna of two Panamanian forests', *J Mammal*, 51 (1970), 473–90
56 Flowerdew, J. R. 'The subcaudal glandular area of *Apodemus sylvaticus*', *J Zool Lond*, 165 (1971), 525–7
57 Formozov, A. N. 'Adaptive modifications of behavior in mammals of the Eurasian steppes', *J Mammal*, 47 (1966), 208–23
58 Freuchen, P. and Salomonsen, F. *The Arctic Year*, 1958
59 Genelly, R. E. 'Ecology of the common mole-rat (*Cryptomys hottentotus*) in Rhodesia', *J Mammal*, 46 (1965), 647–65
60 Gilmore, R. M. 'Cyclic behavior and economic importance of the Ratamuca (*Oryzomys*) in Peru', *J Mammal*, 28 (1947), 231–41
61 Grant, E. C., Mackintosh, J. H. and Lewill, C. J. 'Effects of a visual stimulus on the agonistic behaviour of the golden hamster', *Z Tierpsychol*, 27 (1970), 73–7
62 Grey Owl. *Pilgrims of the wild* (1935)
63 Gruneberg, H. *Genetics of the mouse* (1943)
64 Hanney, P. 'The Muridae of Malawi (Africa: Nyasaland)', *J Zool Lond*, 146 (1965), 577–633
65 Happold, D. C. D. 'Biology of the jerboa, *Jaculus jaculus Butleri* (Rodentia, Dipodidae), in the Sudan', *J Zool Lond*, 151 (1967), 257–75
66 Hatt, R. T. 'Lagomorpha and Rodentia other than Sciuridae, Anomaluridae and Idiuridae, collected by the American Museum Congo Expedition', *Bull Am Mus nat Hist*, 76 (1940), 457–604
67 Herberg, L. J., Pye, J. G. and Blundell, J. E. 'Sex differences in the hypothalamic regulation of food hoarding: hormones versus calories', *Anim Behav*, 20 (1972), 186–91
68 Hershkovitz, P. 'Evolution of neotropical cricetine rodents (Muridae)', *Fieldiana: Zoology*, 46 (1962)
69 Hoare, C. A. 'Reservoir hosts and natural foci of human protozoal infections', *Acta Tropica*, 19 (1962), 281–317
70 Horner, B. E., Taylor, J. M. and Padykula, H. A. 'Food habits and gastric morphology of the grasshopper mouse', *J Mammal*, 45 (1964), 513–35
71 Horner, B. E. and Taylor, J. M. 'Growth and reproductive behavior in the southern grasshopper mouse', *J Mammal*, 49 (1968), 644–60

72 Howard, W. E., Marsh, R. G. and Cole, R. E. 'Food detection by deer mice using olfactory rather than visual clues', *Anim Behav*, 16 (1968), 13–17

73 Hudson, W. H. *The Naturalist in La Plata* (1892)

74 Ingles, L. G. *Mammals of the Pacific states*, Stanford (1965)

75 Ireland, P. H. and Hays, H. A. 'A new method for determining the home range of woodrats', *J Mammal*, 50 (1969), 378–9

76 Irving, L. 'Maintenance of warmth in arctic animals', *Zool Soc Symp*, 13 (1964), 1–14

77 Jahodu, J. C. 'The effect of the lunar cycle on the activity pattern of *Onychomys leucogaster breviauritis*', *J Mammal*, 54 (1973), 544–9

78 Jameson, E. W. 'Patterns of hibernation of captive *Citellus lateralis* and *Eutamias speciosus*', *J Mammal*, 45 (1964), 455–60

79 Jarvis, J. U. M. and Sale, J. B. 'Burrowing and burrow patterns in East African mole-rats *Tachyoryctes*, *Heliophobius* and *Heterocephalus*', *J Zool Lond*, 163 (1971), 451–79

80 Kenagy, G. J. 'Saltbush leaves: Excision of hypersaline tissue by a kangaroo rat', *Science*, 178 (1972), 1094–6

81 Kikkawa, J. 'Movement, activity and distribution of the small rodents *Clethrionomys glareolus* and *Apodemus sylvaticus* in woodland', *J Anim Ecol*, 33 (1964), 259–99

82 King, J. A. 'Social behavior, social organization, and population dynamics in a black-tailed prairie dog town in the Black Hills of South Dakota', *Contrib Lab Vertebrate Biol Univ Mich*, 67 (1955), 1–123

83 Kirmiz, J. P. *Adaptations to desert environment* (1962)

84 Krebs, C. J., Gaines, M., Keller, B., Myers J. and Tamarin, R. 'Population cycles in small rodents', *Science*, 175 (1973)

85 Layne, J. N. 'Tail autonomy in the Florida mouse', *Peromyscus floridanus*', *J Mammal*, 53 (1972), 62–71

86 Linsdale, J. M. *The California ground squirrel*, California (1946)

87 Linsdale, J. M. and Tevis, L. P. *The dusky-footed woodrat*, California (1951)

88 Lish, R. D., Russell, J., Kahler, S. and Hanks, J. 'Regulation of hormonally mediated maternal nest structure in the mouse (*Mus musculus*) as a function of neonatal hormone manipulation', *Anim Behav*, 21 (1973), 296–301

89 Lloyd, H. G. 'Observations on nut selection by a hand-reared grey squirrel (*Sciurus carolinensis*)', *J Zool Lond*, 155 (1968), 240–4

90 Longhurst, W. 'Observations on the ecology of the gunnison prairie dog in Colorado', *J Mammal*, 25 (1944), 24–36

91 Lott, D. F. and Hopwood, J. H. 'Olfactory pregnancy block in mice (*Mus musculus*)', *Anim Behav*, 20 (1972), 263–7

92 Loukashkin, A. 'Giant rat-headed hamster *Cricetulus triton nestor* Thomas of Manchuria', *J Mammal*, 25 (1944), 170–7

93 Marlow, B. J. 'A comparison of the locomotion of two desert-living Australian mammals, *Antechinomys spenceri* (Marsupalia: Dasyuridae) and *Notomys cervinus* (Rodentia: Muridae)', *J Zool Lond*, 157 (1969), 159–67

94 Marsden, W. *The lemming year* (1964)

95 Martin, P. 'Movements and activities of the mountain beaver (*Aplodontia rufa*)', *J Mammal*, 52 (1971), 717–23

96 Maser, C. O. 'Commensal relationship between *Acomys* and *Rousettus*', *J Mammal*, 47 (1966), 153

97 Mayhew, H. *London labour and the London poor* (1861)

98 McNamara, M. C. and Riedesel, M. L. 'Memory and hibernation in *Citellus lateralis*', *Science*, 179 (1973), 92–4

99 McCarley, H. 'Annual cycle, population dynamics and adaptive behavior of *Citellus tridecemlineatus*', *J Mammal*, 47 (1966), 294–316

100 Medway, Lord. *The wild mammals of Malaya* (1969)

101 Mitchell, O. G. 'The supposed role of the gerbil ventral gland in reproduction', *J Mammal*, 48 (1967), 142

102 Morris, D. 'The behaviour of the green acouchi (*Myoprocta pratti*) with special reference to scatter hoarding', *Proc Zool Soc Lond*, 139 (1962), 701–32

103 Morton, M. L. and Tung, H. C. 'Growth and development in the belding ground squirrel (*Spermophilus beldingi beldingi*)', *J Mammal*, 52 (1971), 611–16

104 Mrosovsky, N. 'The adjustable brain of hibernators', *Scientific American* (1968), 110–18

105 Muul, I. 'Day length and food caches', *Natural History*, 74 (1965), 22–7

106 Muul, I. 'Photoperiod and reproduction in flying squirrels, *Glaucomys volans*', *J Mammal*, 50 (1969), 542–9

107 Myers, G. T. and Vaughan, T. A. 'Food habits of the plains pocket gopher in eastern Colorado', *J Mammal*, 45 (1964), 588–98

108 Mykytowycz, R. 'The role of skin glands in mammalian communication', *Communication by Chemical Signals*, New York (1970), 327–60

109 Negus, N. C. and Pinter, A. J. 'Reproductive response of *Microtus montanus* to plants and plant extracts in the diet', *J Mammal*, 45 (1966), 69–87

110 Nevo, E. 'Observations on Israeli populations of the mole rat, *Spalax e. ehrenbergi* Nehring 1899', *Mammalia*, 25 (1961), 127–43
111 Nevo, E. and Amir, E. 'Geographic variation in reproduction and hibernation patterns of the forest dormouse', *J Mammal*, 45 (1964), 69–87
112 Noirot, E. 'Serial order of maternal response in mice', *Anim Behav*, 17 (1969), 547–9
113 Norman, F. I. '*Rattus rattus* population inhabiting colonies of short-tailed shearwater (*Puffinus tenuirostris*) on Big Green Island, Tasmania', *J Zool Lond*, 162 (1970), 493
114 Ognev, S. I. *Mammals of the USSR and adjacent countries*, Vols 5, 6 and 7. Translated from Russian 1964. Israel program for scientific translations (1950)
115 Parkes, A. S. and Bruce, H. M. 'Olfactory stimuli in mammalian reproduction', *Science* (1961), 1049–54
116 Palmer, R. S. *The mammal guide: Mammals of North America north of Mexico*, New York (1954)
117 Pearson, O. P. 'Life history of mountain viscachas in Peru', *J Mammal*, 29 (1948), 345–74
118 Pearson, O. P. 'Mammals in Peru', *Bull Mus Comp Zool* (1951), 117–71
119 Petite, I. *The elderberry tree* (1965)
120 Quanstrom, W. R. 'Flood tolerance in Richardsons ground squirrel', *J Mammal*, 47 (1966), 323
121 Quanstrom, W. R. 'Behavior of Richardsons ground squirrel, *Spermophilus richardsoni richardsoni*', *Anim Behav*, 19 (1971), 646–52
122 Quick, H. F. 'Occurrence of porcupine quills in carnivorous mammals', *J Mammal*, 34 (1953), 257
123 Ralls, K. 'Mammalian scent marking', *Science*, 171 (1971), 443–9
124 Reig, O. 'Ecological notes on the fossorial octodont rodent *Spalacopus cyanus* (Molina)', *J Mammal*, 51 (1970), 592–601
125 Reimer, J. D. and Petros, M. L. 'Breeding structure of the house mouse *Mus musculus*, in a population cage', *J Mammal*, 48 (1967), 88–99
126 Rice, D. W. 'Sexual behavior of tassel-eared squirrels', *J Mammal*, 38 (1957), 129
127 Richens, V. B. 'Notes on the digging activity of a northern pocket gopher', *J Mammal*, 47 (1966), 531–3
128 Rongstad, O. J. 'A life history study of thirteen-lined ground squirrels in Southern Wisconsin', *J Mammal*, 46 (1965), 76–87
129 Ropartz, P. 'L'urine de souris en tant que source odorante re-

sponsable de l'augmentation de l'activite locomotrice', *Rev Comp Animal*, 4 (1967), 71–82

130 Ropartz, P. 'The relation between olfactory stimulation and aggressive behaviour in mice', *Anim Behav*, 16 (1968), 97–100

131 Rosevear, D. R. *The rodents of West Africa* (1969)

132 Rowe, F. P. 'The response of wild house mice (*Mus musculus*) to live traps marked by their own and by a foreign odour', *J Zoo Lond*, 162 (1970), 517–20

133 Rowe, F. P. and Taylor, E. J. 'Numbers of harvest mice (*Micromys minutus*) in corn ricks', *Proc Zool Soc Lond*, 142 (1964), 181–5

134 Ruffer, D. G. 'Sexual behavior of the northern grasshopper mouse (*Onychomys leucogaster*)', *Anim Behav*, 13 (1965), 447–52

135 Sales, G. D. 'Ultra-sound and aggressive behaviour in rats and other small mammals', *Anim Behav*, 20 (1972), 88–100

136 Sanderson, I. T. 'The mammals of the North Cameroon forest area being the results of the Percy Sladen expedition to the Mamfe division of the British Cameroons', *Trans Zool Soc Lond*, 24 (1940), 623–725

137 Schmidt-Nielsen, K. *Desert animals* (1964)

138 Sewell, G. D. 'Ultrasonic communication in rodents', *Nature*, 227 (1970), 410

139 Shadle, A. R., Smelzer, M. and Metz, M. 'Sex reactions of the porcupine (*Erithizon d. dorsatum*) before and after copulation', *J Mammal*, 27 (1946), 116–21

140 Shadle, A. R. 'Removal of foreign quills by porcupines', *J Mammal*, 36 (1955), 463–5

141 Shapiro, J. 'Ecological and life history notes on the porcupine in the Adirondacks', *J Mammal*, 30 (1949), 247–57

142 Sharp, H. F. 'Food ecology of the rice rat, *Oryzomys palustris*, in a Georgia salt marsh', *J Mammal*, 48 (1967), 557–63

143 Sheets, R. G., Linder, R. L. and Dahlgren, R. B. 'Burrow systems of prairie dogs in South Dakota', *J Mammal*, 52 (1971), 451–3

144 Shellhammer, H. S. 'Cone-cutting activities of Douglas squirrels in sequoia groves', *J Mammal*, 47 (1966), 525–6

145 Shkolinck, A. and Borut, A. 'Temperature and water retention in two species of spiny mice (*Acomys*)', *J Mammal*, 50 (1969), 245–55

146 Shorten, M. *Squirrels* (1954)

147 Shrewsbury, J. F. D. *A history of bubonic plague in the British Isles* (1970)

148 Smith, R. E. 'Natural history of the prairie dog in Kansas', *Univ Kansas Misc Publ*, 49 (1967)

149 Sokolov, W. and Skurat, L. 'A specific midventral gland in gerbils', *Nature*, 211 (1966), 546–7

150 Southern, H. N. *Handbook of British Mammals* (1964)

151 Sowls, L. K. 'The Franklin ground squirrel, *Citellus franklinii* (Sabine), and its relationship to nesting ducks', *J Mammal*, 29 (1948), 113–37

152 Spanuth, H. *Der Rattenfanger von Hameln*, Hameln (1951)

153 Stephenson, A. B. 'Temperature within a beaver lodge in winter', *J Mammal*, 50 (1969), 134–6

154 Stoddard, D. M. 'A note on the food of the Norway lemming', *J Zool Lond*, 151 (1967), 211–13

155 Stoddard, D. M. 'The lateral scent organs of *Arvicola terrestris* (Rodentia: microtinae)', *J Zool Lond*, 166 (1972), 49–54

156 Swanson, H. 'The consequences of overpopulation', *New Scientist* (1973), 190–2

157 Taylor, J. 'The use of marking points by grey squirrels', *J Zool Lond*, 155 (1968), 246–7

158 Thiessen, D. D., Blum, S. and Lindzey, G. 'A scent marking response associated with the ventral sebaceous gland of the mongolian gerbil (*Meriones unguiculatus*)', *Anim Behav*, 18 (1970), 26–30

159 Thiessen, D. D., Lindzey, G., Blum, S. and Wallace, P. 'Social interactions and scent marking in the mongolian gerbil (*Meriones unguiculatus*)', *Anim Behav*, 19 (1971), 505–13

160 Tittenor, A. M. 'Red squirrel dreys', *J Zool Lond*, 162 (1970), 528–33

161 Thorington, R. W. 'Lability of tail length of the white-footed mouse, *Peromyscus leucopus noveboracensis*', *J Mammal*, 51 (1970), 52–9

162 Troughton, E. *Furred mammals of Australia* (1941)

163 Turkowski, F. J. 'Resistance of the roundtail ground squirrel (*Spermophilus tereticaudus*) to venom of the scorpion (*Centruroides sculpturatus*)', *J Mammal*, 50 (1969), 160–1

164 Valenta, J. G. and Rigby, M. K. 'Discrimination of the odour of stressed rats', *Science*, 161 (1968), 599–601

165 Van de Graff, K. M. and Balder, R. P. 'Importance of green vegetation for reproduction in the kangaroo rat, *Dipodomys merriami*', *J Mammal*, 54 (1973), 509–12

166 Vaughan, T. A. 'Food-handling and grooming behavior in the plains pocket gopher', *J Mammal*, 47 (1966), 123–33

167 Walker, E. P. *Mammals of the world*, Baltimore (1964)

168 Wallace, A. F. and Lathbury, V. L. 'Culture and the beaver', *Natural History*, 77 (9) (1968), 58–65

169 Warwick, T. 'A contribution to the ecology of the musk rat (*Ondatra zibethica*) in the British Isles', *Proc Zool Soc Lond*, 110 (1940), 165–201
170 Webster, D. B. 'Ears of *Dipodomys*', *Natural History*, 74 (1965), 27–33
171 Whittaker, J. O. 'Food of *Mus musculus*, *Peromyscus maniculatus bairdii* and *Peromyscus leucopus* in Vigo County Indiana', *J Mammal*, 47 (1966), 473–86
172 Whitten, W. K., Bronson, F. and Greenstein, J. 'Estrus-inducing pheromones of male mice: Transport by movement of air', *Science*, 161 (1968), 584–6
173 Whitten, W. K. and Bronson, F. H. 'The role of pheromones in mammalian reproduction', *Communication by chemical signals*, New York (1970), 309–25
174 Wilsson, L. *My beaver colony* (1969)
175 Zeuner, F. E. *A history of domesticated animals* (1963)
176 Zinsser, H. *Rats, lice and history* (1934)
177 Zumpt, I. 'The ground squirrel', *Afr wild Life*, 24 (1970), 115–21

ACKNOWLEDGEMENTS

I should like to record my thanks to the following authors for allowing me to reproduce figures from their publications; to Prof G. N. Cameron (Fig 32), David Happold (Fig 15) and Dr O. J. Rongstad (Fig 33). The Zoological Society of London kindly gave permission for the reproduction of Fig 15, and the American Society of Mammalogists for Figs 32, 33. I am also indebted to those who helped in supplying photographs and information, especially Mr C. E. V. Fullaway of Hudson's Bay and Annings Ltd, Mrs Virginia Lathbury and Mr A. J. Parkin. Since the beginning of my interest in rodents, I have always received friendly advice and assistance from Dr G. B. Corbet, Mr J. E. Hill and staff of the Mammal Section of the British Museum. Lastly, I am indebted to my wife who has patiently cared for many of the subjects of this book, and has often suffered considerable discomforts while accompanying me on their trail.

INDEX